# Beekeeping
*in the*
# End Times

LARISA JAŠAREVIĆ

# Beekeeping in the End Times

INDIANA UNIVERSITY PRESS

This book is a publication of

Indiana University Press
Office of Scholarly Publishing
Herman B Wells Library 350
1320 East 10th Street
Bloomington, Indiana 47405 USA

iupress.org

© 2024 by Larisa Jašarević

All rights reserved
No part of this book may be reproduced or utilized in any form or by any means, electronic or mechanical, including photocopying and recording, or by any information storage and retrieval system, without permission in writing from the publisher. The paper used in this publication meets the minimum requirements of the American National Standard for Information Sciences—Permanence of Paper for Printed Library Materials, ANSI Z39.48–1992.

Manufactured in the United States of America

First Printing 2024

Cataloging Information is available from the Library of Congress.

ISBN 978-0-253-06811-8 (hardcover)
ISBN 978-0-253-06812-5 (paperback)
ISBN 978-0-253-06813-2 (ebook)

# CONTENTS

*Acknowledgments*  *vii*

1. Are the Bees Still Swarming? A Tale of Two Angels  *1*
2. Honey's on the Wheels: Beekeepers' Prayers  *50*
3. Planting on the Eve of Apocalypse: The Prophet's Advice  *85*
4. Near-End Ecology: A Devil at Heart  *120*
5. The End: God's Promise  *158*

*Bibliography*  *185*

*Index*  *195*

# ACKNOWLEDGMENTS

Truly, the angels lower their wings, taking pleasure in the seeker of knowledge. The inhabitants of the heavens and Earth, even the fish from the depths of the water, seek forgiveness for the scholar.

So goes a saying by the Prophet of Islam. This prophetic saying (hadith) praises the quest for knowledge in a curious way, as a pursuit that makes a human seeker of concern to the angels and the animals.

Scholarship, in the widest sense of seeking to know, whether by reading a text or doing long-term research among the things of the world, is somehow a trial with ramifications for the skies and the earth, the hadith suggests. We learn that the angels relish the sight of a knowledge seeker. Some versions of this prophetic saying and their commentaries suggest that angels lower their wings out of humbleness before the seekers or to envelop them protectively. My eyes, however, are just as caught by the second part of the hadith, which intimates a more somber, indeed cautious, multispecies response to human pursuits of knowledge. As the attention of the hadith widens to become ecological, human knowledge is no longer primarily an object of admiration or pleasure but rather something possibly pernicious, a liability that makes the fish in the depths of the sea wary. Nonetheless, the fish respond with compassion because compassion is at the root of existence and also, perhaps, because all beings know something about the grave implications of knowledge that humans must strive to find out. Perhaps it is the residents of Earth, the ones most vulnerable to human action (which, by default, enacts a knowledge of sorts), that rally the heavens, too, in invoking God's forgiveness of humans. The prophetic saying gives insight into an eco-cosmological Islamic perspective on humans as beings defined by utter accountability.

The writing and research that have gone into this book have brought me much pleasure in learning and just as much worry about what I have discovered. When it comes to understanding and sharing the findings, my limitations have nagged me from start to finish. Luckily, in the course of the long endeavor that this book required, I have not been left to my own means.

Several institutions earn my hearty thanks for their support throughout the project. Foremost among them is the American Council of Learned Societies (ACLS), which funded the field research from 2016 to 2017. A fellowship through the ACLS/Luce Program in Religion, Journalism & International Affairs saw me through the first year of writing. What is more, ACLS's wonderful staff, whether they know it or not, have helped launch the documentary filmmaking on the subject of this research—a quietly held hope I have long had.

Just as wonderfully supportive was the Wenner-Gren Foundation, which cofunded the field research from 2016 through 2018. Wenner-Gren's Fejos Postdoctoral Fellowship for the Ethnographic Filmmaking has helped make this book far richer than I could have imagined. While my sister Azra Jašarević and I excitedly embarked on the film production, the experience of thinking with the camera, "keeping the plot in mind," as Azra insisted, has deeply affected my writing (it is a work in progress). What is more, the footage we collected in 2021 contributed essential material to the book's story of climate change effects on local honey ecologies, which by then were becoming more palpable on the honey routes and more succinctly vocalized in beekeepers' commentaries.

Most recently, a fellowship with the Independent Social Research Foundation through 2022 has enabled me to finish this book and begin the film's postproduction. These academic institutions, with their staff devoted to the fellows, anonymous reviewers who are generous with their time and talent, and patrons committed to supporting scholarship through dire times on our planet, financially and otherwise, count among the blessings.

Students, colleagues, and friends at the University of Chicago get credit for helping me initially articulate the rationale for this project. Students who joined classes I taught between 2015 and 2018 made me only more eager to try to live differently, with and beyond the texts we cherished. I am also thankful to the Max Planck Institute for the History of Science in Berlin, which welcomed me as a visiting scholar through 2022.

Jennika Baines, my editor with Indiana University Press from 2018 to 2021, gets all the credit for being enthusiastic about the project, being gracious with her advice, and steering my writing in a particular direction when she said that she hoped I would write a book her mother would appreciate. I tried. Allison Chaplin competently directed the manuscript through the last revisions. Nancy Lila Lightfoot saw the book through the production and shared precious news from the insect haven in her backyard garden.

I thank the anonymous reviewer—a beekeeper—for the Press for being frank about the kinds of materials that sustain attention of beekeepers. I hope such readers will find here an appeal to voice their field observations as crucial contributions to the emergent inquiry into the climate change effects on the green planet.

My deepest gratitude goes to Anna Tsing, who read the manuscript closely and generously more than once. Anna's hopes for the project encouraged me when I was deeply uncertain how to write this book. Her readings helped me grasp the implications of my materials and have emboldened my arguments. Anna's support brings to mind another hadith, which says that even the ant in his hole and the fish, send blessings on the one who teaches people what is good. Bless you.

One long conversation with Reza Aslan convinced me to give a second thought to the fiery species that I kept on the margins of my field notes. Reza may not be aware of it, but I owe the stride of chapter 4 to him. The way he carries his superb scholarship across multiple media remains deeply inspiring.

Caryn O'Connell, a friend, a bard, and an editor without peer, worked with me through the very first versions of the book through

2019, not least during her visit to Bosnia. Our cross-country journey was hilarious and not without its awesome moments, in the old sense of that word. Larisa Kurtović, a fellow anthropologist, was involved with this project over the years, the way true friends are: by being concerned with my well-being as much as with the vagaries of writing.

In Bosnia, my sister Azra often joined me on field research trips even before we started filming. Azra also got hooked on bees and began dreading with me the near future of climate change. Sharing both apiary tasks and unease makes the day-to-day more manageable and far more fun. My sister Mirza and her family provided all the commonsense advice and practical support one needs to function while writing while Mirza made filming arrangements on sites that are off limits to the first-time filmmakers. I cannot thank them enough. Pepe, the cat, is the sort of companion every library and in-home production studio needs to help keep human priorities in perspective.

And finally, my mother, Zumreta, who keeps up our piece of land almost single-handedly while I am occupied with writing. She grows and cooks our food, keeps the orchard and countless flower gardens for our resident insects and birds, brings me firewood when it is cold, and has helped me catch every swarm. Over dinners, she patiently heard out numerous drafts of this book. The blissful Garden is under your mother's feet, a hadith, says, and I pray every step brings her closer to *al Jannah.*

The only cure for ignorance is asking questions, the Prophet is said to have said. I am utterly in debt to the many beekeepers and bee lovers across Bosnia and Herzegovina who have indulged my questions, hosted me at their apiaries over the years, gifted me swarms, hives, and jars of precious honey. The Velagić family, Enver and Adil, continue to mentor me in apiculture. May your honeybees flourish through the rough times.

The generosity of Enes Karić, a professor at the Faculty of Islamic Studies, University of Sarajevo, must be legendary in present-day Bosnia and Herzegovina. I thank him for being supportive of my research and writing. The staff of the Gazi Husrev-Beg Library in Sarajevo assisted me with gathering records on the history of apiculture in Bosnia and Herzegovina. Professor Behija Dukić at the University of Veterinary Science in Sarajevo kindly introduced me to many beekeepers. Her interest in the therapeutic properties of hive substances made me better appreciate the implications of artificial feeding for the species whose artifacts hold fast the attention of numerous clinical and lab researchers around the world.

Humans aside, the honeybees have thoroughly shifted my relationship with the world. I remain truly moved and amazed by them. Out of gratitude and compassion, I wish to care for them and their companion plants for as long as I can, at least within the bounds of our land or our village. I hope they invoke forgiveness on my behalf, for this poor human surely needs it, even before Jibrīl tends to my lot on the divine scale, his six hundred wings rolled up like a keen accountant's sleeves.

Local Sufis have taught me what angels know very well; namely, that knowledge, though praiseworthy, will only get you so far. At the furthest limits of the knowable, by the shade of the Lote Tree beyond which even the greatest of angels, the loyal Jibrīl, pure intellect, cannot pass but halts sensibly. Another step toward the divine would enflame his luminous wings. Many dervishes and their guides are drawn further, because they cannot resist. Anyone who has ever been head-over-heels in love will understand something about the urge to go further, that seekers of God feel. *Love, friendship,* and *adoration* are separate English words for the feelings that go into the same brew within dervish circles. Love (+ friendship + adoration) is another means of knowing.

Shaykh Ayne, a Naqshbandi elder, graciously agreed to teach me even though the task burdened his frail frame and our companionship inconvenienced him. Out of wisdom, he taught me first of all what I did not want to learn. Had it not been for his gentleness and perseverance, I would have walked away from those introductory lessons of the first few years. I am eternally indebted to Shayk Ayne, but I trust that he has already been lavishly repaid for the trouble, in the Barzakh. Inshallah. Had it not been for Shaykh Ayne's closest guides, the wise and witty Shaykh Mustafa, and the luminous, dearly loved Shaykh Halid, I would have been stuck when Shaykh Ayne passed on. They make true a line that Ibn al-'Arabī bestowed in his book of prayers, that closeness is at the heart of distance.

I thank my uncle Faruk, who understood deeply what I was trying to write and for whom I did not need to put moods into words. I wish love and peace on him. The dwellers of Barzakh, the vast land that the tradition paints between the worlds where the wildest opposites reconcile, can have both at once, at last.

I thank all other Shaykhs and dervishes who have gifted me their time and insights. I single out one: Zejd. He is teaching me the ins and outs of unconditional gratitude. If I had any wings, I would lower them for him. Enough said.

# Beekeeping in the End Times

# − 1 −

## *Are the Bees Still Swarming?*
### A Tale of Two Angels

Two angels in the near heavens open their eyes once every hundred years. "Are the bees still swarming?" they ask. Their fellow angels who keep watch on Earth respond: "The bees are still swarming." "What about the sheep? Are they still lambing?" the two angels wonder aloud. "The sheep are still lambing," the other angels reassure them. "And the fish? Are they still spawning?" the two ask finally. "The fish are still spawning," the watchful angels kindly reply. "Well, then," the two angels conclude their inquiry: "the End [*Kijamet*, pronounced *Keeyamet*] is not just yet," and they close their eyes for another spell of silent invocation.

This book is written in the loaded, uneasy moment of the "not just yet," a moment made present for listeners by the tale of the two angels each time it is retold in present-day Bosnia and Herzegovina (BiH). That the End is looming is no news to Bosnian Muslims. The finite nature of Earth, the cosmos, and all living things is the basic lesson of the Islamic sacred texts, and the acceptance of personal death is a way in which Muslims live in Islam.

And yet the power of this wisdom tale is such that with each retelling, its warning sounds fresh, and the End the angels await looms closer still and more daunting. The tale seeks to awake humans distracted from matters of finitude and forgetful of the solemn divine reminder that one day the world will end. In a few short moments, it teaches remembrance of this world, its importance and impermanence, and its utter frailty and sacredness, since the world both reveals and veils God and subsists on God's attention. Remembrance is a practice essential to being a Muslim. Forgetfulness, conversely, is the road to ruin. *And be not like those who forget God, so He makes them forget themselves. Those are the defiant.*[1]

The story also rehearses the order of things: a cosmology in which Earth is tied to the heavens, angels are mindful of the animals, and God's invocation is every being's vocation. Addressed to humans, the tale omits explicit mention of us while all along hinting at the very human, ruinous dispositions at work in the world and our species' grave responsibilities in the face of the coming End.

The appeal of the wisdom tale lies in how it reminds us of the End. It invokes tenets of the Scripture and the rich tradition of its commentaries, but it points out into the living world. It raises questions. The angels themselves can do no better than inquire and investigate. If you want to know the state of Earth, pay attention to animals, the tale suggests. And the angels' first concern is with the honeybees. Are they swarming?

For years now, I have been paying attention to the beekeepers and the honeybees as they forage for nectar and pollen through the deranged seasons, strange weather, and changing ecologies of BiH. From 2014 to 2019, I conducted what anthropologists call an *ethnography*: a field-based observation and hands-on, close involvement with cultural practices. To this day, I keep bees.

The following pages convey the results of my investigations, gathered on the trail of seasoned apiculturists who are traveling familiar routes through environments traditionally known for copious honey flow. These environments are rapidly changing. With the materials on hand, I cannot tell the climate future of honeybees or pollinators at large. Nonetheless, the range, form, and quality of alterations I have recorded add a precious and missing record of the ongoing impacts of climate change on interspecies relations between plants, bees, and humans. This record is important given how little is known about the effects of global warming on what biologists call *pollination ecologies*—the blooming field of interactions between plants and insects that keep the earth green and alive and, in turn, replenish the atmosphere under the local skies. The honey ecologies that preoccupy local beekeepers presume a still wider range of signs, meanings, and vital attractions that keep this world thriving. The waning of honey appears to them as a sign of the times.

By sharing my notes on how local beekeeping fares through changing climates, I also hope to diversify the questions and concerns about what is going on at present. The beekeepers themselves often wonder aloud on this topic. Their questions remain open as they try out old and new strategies to sustain the bees and to pursue prized honey and pollen. Whether such strategies entail more agile traveling and forage forecasting or tireless, diversified planting to improve the odds of honey and pollen harvest, they show well the extent to which resourceful responses are both fostered and foreclosed by the very local grounds where people act.

When one raises unnerving questions in specific places amid the clutter of ecological-historical circumstances that spell out limits of action and depth of attachments, complex and daunting questions, such as those about the effects of climate change, take on a loamy meaning. Values that are sought in the here and now by particular people and bonds presumed and tended among humans, animals, and plants are all too rich to be assimilated into formal terms and assessments of climate risk.

Unease may be the first step in grasping the "not quite yet" that has a hold of us, whether or not we are awake to the possibility that we are living in the end times. Catastrophic thinking, however, can be darkly seductive, which makes it rather unpopular among scholars, scientists, and public environmentalists. Apocalypticism is the most blatant form of bleak expectations and the most controversial. Describing something as "doomsday thinking" can lead the listener to belittle its logic and relevance. The worst trouble with the invocation of the looming apocalypse, however, is not so much that it seriously considers the world's imminent collapse but that it too often hastens to resignation and the dead end. This book, therefore, is neither apocalyptic in the usual sense of the word nor is it averse to the apocalyptic traditions that deeply inform modern Western imagination, religious as much as secular. It brings forth and foregrounds Islamic eschatological lore: teachings, contemplations, and mindfulness about living in the light of imminent death and the world's ultimate ruination.

Eschatology, as such, is the fiber of Islamic tradition. It informs all dimensions of faith practice and religious thinking to animate projects as diverse as charitable giving, child-rearing, and political or militant mobilizing. Political and militant Muslim eschatologies, with their grave speculations, gory imagery, or overly simplistic rendering of the good and the evil, here and in the afterlife, tend to capture media and scholarly attention. This book, by contrast, points to the profuse but more subtle relevance of Islamic eschatology in canonical texts, in the living spiritual tradition of Sufism, and in contemporary Muslim lives.[2]

Eschatology makes sense of life. Highlighting the finitude of all things, it revalues each life and the sheer effort of living well. It teaches how to live. The pages of the Qur'an, which Muslims read devoutly, recite in their daily prayers, and retell or cite to each other for comfort or as food for thought, are eschatological more often than they are not. Read with me, for instance, the following lines: *Death throes will come, in truth; that's what you've been avoiding. And the Horn will blast—it will be the promised Day.*[3]

But the eschatological tone itself is all important for the divine message to carry well, and that tone always blends warnings with comfort and cheer. The Qur'an, too, is the message of the "not just yet," and the news of divine promise it delivers is at once sober and sweet. *O Prophet, truly we have sent you as a witness and a bringer of glad tidings and as a warner.*[4]

A dear friend, Zejd, a Sufi dervish with formal training in Islamic theology, once wrote a story to help me convey what eschatology does. Here is Zejd's story: A man walked far to seek out the blessed Prophet of Islam. He had heard the revealed message and taken it to heart, but he wanted much more precision on the subject of this looming End: What it is like? When it will be? and such. The Prophet heard him out and smiled. Instead of soothing the man's fears or furnishing the missing details, he simply asked in turn: "And what have you prepared for It?" The Prophet, in other words, shifted the man's perfectly reasonable questions to what counts: There's the solemn certainty of the End—*when* the Hour comes, God knows when—and you and your doings will be seen in its light. Doing well in the present, the rest of the book shows, entails doing what foresees and forestalls the coming apocalypse.

It is such practical-minded eschatological activism that concerns me in this book. This sort of practice, however, is closely tied to contemplation, not least because these grave subject matters encourage pondering profound questions such as the nature and meaning of existence. Thinking about life and death, in other words, entertains metaphysical questions about what is real and what really matters.

Metaphysics, meaning foundational ideas about the world we live and die in, may seem out of place in mundane settings such as apiaries or gardens or a distraction at the height of an ecological crisis. On the contrary, such questions are already implied as we go about our daily business and, although often unstated, bear powerfully on how we handle ourselves and approach others, including plants and animals, and even how we relate to the weather day to day. Metaphysics is central to the stories we tell each other about the state of our world and its current course toward ruination or chances for remediation.

That the biosphere is going to rack and ruin also decidedly shows a metaphysical crisis of the global, late-modern cultural economy as devastating as the effects of the fossil fuels that are spinning its wheels. The deepest metaphysical questions are most readily glossed over as we make arguments for and against business as usual, voice our hopes about livable futures, and consider proposals for piecemeal adaptations or radical alternatives: Who or what are we? What is this place, Earth? And what the heck are we doing here to begin with?

The wisdom tale about the two angels offers poignant clues about basic Islamic metaphysics. Whereas in the hadith, the Prophet's question that Zejd related to me—"And what have you prepared for It?"—focuses on humans and emphasizes the importance of human deeds for living and dying well in God's good pleasure, the angel's story slightly shifts the perspective.

Told from the angels' point of view, the wisdom tale foregrounds vital relationships between Earth and the heavens in which humans play a particular part. Angels are deeply interested in the animals, and their repeatedly stated concerns and watchfulness are expressions of divine care for animal flourishing. The local listeners are presumed to know that the suspected threats to animals are due to people making a mess of things on Earth or, in Qur'anic terms, doing *fasād*.⁵ The angels' story further teaches that the fate of the world is inseparable from animal well-being. This particular wisdom tale (*hikaja*, from the Arabic *ḥikāya*, meaning "story") is a clever folk commentary on the Islamic textual tradition that speaks of God as the Nurturer or Lord of the Worlds.

Tendency to focus on the human, and man in particular, as the principle agent in the narratives of this world's history, are as strong in modern Islamic thought as they are in modern projects the world over, but that sort of humanism whittles down the teeming worlds of other beings that God invokes in the revealed words. Divine self-descriptions as the Nurturer and the Lord of the Worlds in the Qu'ran presume more than a human cosmos while verses—and their commentaries—stress that everything, not just humans, subsists on divine care and attention. Moreover, the world itself is the divine self-revelation that is ongoing, as the organic, sensuous signs are lavished upon the cosmos at large and grasped, piecemeal, by each existent according to its singular disposition and species-specific receptivity. Everything in cosmos responds to the divine revelation with praise. Since humans are more ambivalent, defiant, or indifferent—indeed, this is the brunt of our species exceptionalism—Qur'an appeals to the humans.

Put together, the two stories—the angels' and Zejd's—suggest that preparing for the Hour entails doing what is necessary for personal salvation, but that such doing requires, by default, fostering conditions for swarming, lambing, and spawning. Put differently, forestalling the apocalypse presumes that humans give earthly beings their due, honoring God's interest in and presence within other divine subjects. Islamic eschatology is irreducibly ecologically minded; it is, in a word, an eco-eschatology.

At the start of my study, I knew very little about bees. I knew that their honey was praised in the region for its nutritional and medicinal qualities and that people casually described honeybees as divinely inspired. It was obvious to me that the apiculture in BiH did not share the same load of predicaments that overwhelm commercial pollinators in North America and Northern Europe, especially given the small scale of the local operations and honey gathering that preoccupy the apiarists. What I knew and had read by then, however, did not prepare me for appreciating the bees' essential relationship with the weather.

## *The May Storms of 2014*

The rain has finally stopped. I step into the sunlit terrace to take in the damage. It has been raining so relentlessly for so long that our steep piece of land in the mountains has turned into a web of creeks rushing from high up down into the valleys far below, overwhelming the river's watershed in the lowlands with rocks, wood, silt, and debris.

Our flower gardens are sunken, vegetable lots erased. The house was kept dry by makeshift channels dug out in the middle of one long night with pick-shovels, in emergency mode, by three pairs of hands—my sister's, our father's, and my own. Our mother, recovering from a broken leg, was inside with the cat and otherwise employed with prayers, since our tools could only go so far in keeping us all safe. The edges of our land and the fields had already been furrowed by a creeping landslide, and it was only the many trees in our orchard, like troupes of rooted angels, that kept us firmly grounded in the soggy soil.

Stuck inside in the previous days and waiting for the skies to clear, my frustration grew as fast as mold, giving way to dread as I seriously began to wonder whether the rain would ever stop. The old customs of weather—sunshine after rain—could no longer be counted on. On days when the skies cleared, we hurried outdoors to get things done, but the clement spells did not last long: clouds soon gathered, and rainstorms like we had never seen came down like bleak curtains over the familiar world we knew, loved, and feared we might have lost.

On the face of things, the summer-long bad weather affecting our village was not a disaster. At least, it did not register as such in the local, let alone global, media, which was focused on disaster relief and post-disaster recovery in the regions of BiH that had flooded earlier in May following a cyclonic storm. During that storm, three days of extreme rainfall deluged the industrial and agricultural areas located across the watershed of central and northeastern Bosnia. The Sava River, which skirts the country's northern border, peaked with flash floods, its level rising to flood threats faster than anyone could remember.

It did not help that the national river infrastructure was poor due to war damage and peacetime neglect, that there were no early-warning systems in place, or that the river management was inadequately co-ordinated between the administrative units that make up the deeply divided Bosnian state. Sava River management also requires international cooperation since it courses through three other nations: Slovenia, Serbia, and Croatia, but the peacetime relations of these former Yugoslav states remain tense since the breakup of socialist Yugoslavia and the 1990s wars that followed.

Once the rain stopped, foreign relief agencies and self-organized groups of Bosnian citizens poured in to help the displaced and stranded people with emergency aid and to initiate cleanup and recovery projects. Come June, the natural disaster was technically over. It was not the first episode of extreme weather in the region, but the fierceness of the storm and the scale of devastation made the May 2014 floods the watershed event in conversations on global climate change threats to the Balkans region. At least it became such in the statements issued by government officials, United Nations (UN) agencies, the World Bank, and international donors.

On the ground, a different conversation about disaster circulated. "This is worse than war," I heard people in the village say. Referring to the 1990s conflict still fresh in their memory, my neighbors compared the extreme weather event with warfare and found the storm and ongoing downpours even more unsettling.

While the war raged through Bosnia in the 1990s, civilians fortunate enough to have access to land, however small, grew food, foraged, and found comfort in nature. Even the smallest green plots between parking lots were cultivated, while urbanites raided the suburbs, thickets, and forests for wild fruit, berries, mushrooms, herbs, and edible weeds. The extreme weather, however, made the most elemental media of life hostile. Stuck indoors for much of the summer, we felt as if we had been banished from the very earth and atmosphere.

Kijamet, as invoked by the two angels in the wisdom tale, was another way people described the weather catastrophe.

### *A Surprise Swarm*

Sitting on the terrace of our family village home during a sunlit spell at noontime late in June, I am taken by the restored beauty of our land. The paths and gardens had been wrecked, but the damage seems manageable, especially as the clouds and the monochrome sheets of rain lift and the landscape in view emerges, radiant. Mists rise from the warming earth, crowning the fruit trees of our unruly orchard garden with fine, fading halos.

All of a sudden, I feel something stirring about the garden. The feeling is strong. Neither sensuous nor visceral, something is turning my whole person outward, making me curious, expectant, for what I do not know. I stand up, draw to the edge of the terrace, scan the surroundings, hold my breath, and listen anxiously. Then, finally, I see it. There, by the plum tree, a bee swarm is in the making.

Tens of thousands of luminous wings catch the light, shimmering with the energy of an insect call to gather the dispersed collective. Their buzzing resonates until every branch in the orchard seems to join the honeybees' invocation. The very air swirls as the bees circle above the crown of the tree, then alight, one by one, forming a cluster. I quickly jump from the terrace and head toward the tree cautiously, trembling with an unfamiliar excitement as if my nerves had become insective, more numerous and more reactive, in the inner hive of feeling. I stand below them, awed.

In the lull between storms, this swarm has taken off from a hive in our neighbor's apiary. She is a traditional beekeeper and lets her bees swarm at will every season, except that she follows the advance signs of swarming carefully and then waits. When the bees take off, she is ready with implements, spells, and invocations to persuade the swarm to land. But this time, the bad weather has taken her off guard—she had not expected the bees to swarm amid lasting downpours, nor had she noticed any signals of their intentions to swarm. That morning, the whole family had gone to their far fields to collect hay for their dairy cow before the next shower.

Bees have their own way of forecasting weather. They run their hive activities in tune with daily and seasonal shifts in the elements, which profoundly determine the life cycles and secreting moods of the plants they feed on and pollinate. Seasoned local beekeepers learn to foretell the local weather by watching their bees.

I took it as a good sign that this surprise swarm was in the making, although the beekeeper had doubted swarm's likelihood and the public weather forecast had anticipated no fair skies in the near future. But the bees knew. They had been patiently reshuffling their nest through the long-lasting storm, and as soon as the skies cleared for a spell, they burst out, full of faith.

I had just started researching local beekeeping that June. The enduring rainstorms had mostly ruined my field research plans by keeping me and the beekeepers from the apiaries.

Had it been otherwise, though, I would have put the May storm and the bad weather only too readily behind me, to focus, instead, on more pressing research questions I originally intended to study. The relentless torrents seized my attention, instead. The sudden visit of the swarm, on the other hand, recharged my wonder about honeybees.

### *A Swarm Spell, A Prayer*

Once the swarm settles, I rush to get my neighbor, Kada, the beekeeper. I find her on the village outskirts, busy building a haystack with her four children and father-in-law while Bobby, their mutt, runs around the field all wound up. Ever since her husband's small trade business failed and he joined the migrant workforce in Slovenia to pay off debts, Kada, in her midthirties, has been handling ten hives on top of many other duties in the homestead.

She remembers growing up with bees in her grandfather's bee yard. She could always call on her father for advice or consult two retired beekeepers in the village. Her father-in-law helps reluctantly, as he is afraid of stings. But mostly, Kada improvises along the way, trusting the bees to do what they know best.

Her hive management is low-key. In preparation for winter, she treats the bees against varroa mites, using either organic or synthetic miticides. She also bundles up the hives with layers of newspaper and grocery store fliers; the colorful shreds that peek from below the lids through winter look as if the hive tops are sprouting in advance of spring. She aids the bees through winter's dearth with food supplements and, when the flow is good, harvests honey.

The early summer routine of catching the swarms is demanding and thrilling. The swarms that are not caught soon after they exit are likely to alight high in the tree crowns. We have often watched Kada balance, barefoot, some fifteen meters above the ground on frail branches of old, local species of apples and pears, working out strategies to shake off the cluster into a skep, laid out with bunches of fragrant lemon balms that attract bees. Her father-in-law, fully dressed in a protective suit and wearing two veiled hats one on top of the other, typically frets below: "She will break her neck!" But unless they are caught while high up in the trees, the bees could soar higher, "head for the forest," and be gone forever. The hassle at the apiary notwithstanding, Kada could never give up the bees, she says, once she has "learned to love them."

At the news of a runaway swarm, Kada drops everything. Soon enough, with one retired beekeeper in tow, her hands gloved and her waist-long blond braids tucked in a veiled hat, she props a stepladder against the lean plum tree. She climbs. Holding a traditional conical wicker skep skyward so it gapes open just below the cluster, she gives the branch a determined shake. A sudden, massive buzz arises; the bee cluster drops and, missing the skep, breaks like a ripe fruit onto the white sheet spread out below around a spare skep, the sheet shimmering with sprinkled sugar.

Hours later, most of the bees have walked into the offered wicker nest. Kada gently wraps the sheet around the skep and carries it to her apiary just steps outside her house where corn and vegetable patches mingle with the orchard. She shakes the bees off into an empty bee box and closes the lid. *Bismillāhir Rāhmānir Rāhīm*, she whispers. In the Name of God, the Compassionate, the Merciful, the bees have been offered a new home, and only by God's leave will they settle in and stay.

It is from this spot that, in all other early summer seasons, Kada has kept an eye on the boxes. Watching what the bees are doing at the hive entrances and listening to how the hives sound, she knows how to forecast swarms. The swarm's advance is usually preceded by a familiar apian negotiation. The honeybee queen, known locally as *matica*, a bee mother, issues an insistent, high-pitch rallying call: *poot-poot-poot*, to which the resident bees respond with a sonorous *kut-kut-kut*. To speakers of the local language, Bosnian-Serbian-Croatian (BSC for short), *poot* sounds like "trip" (*put*), and the chant of keen responses sounds like a collective question—"when?" (*kad?*). These are the near signs of departure, and yet patience, watchfulness, and luck are needed to catch the exact moment when the swarm is cast.

When that happens, Kada and her kids run up to the hive to persuade the bees to stay with them. She points a water hose skyward, softening the strong stream with her fingers into a sprinkle while pleading and spell-making with the swirling streak of insects: *Sjedi majka, sjedi majka* ("Sit down, mother, sit down, mother"). Her older son plucks fistfuls of the apiary grass and tosses it up in the air, joining his mother's bidding in his changing teenage voice: "Sit down, mother!" His younger sister quickly squats, taps her fingers lightly at the ground, then jumps up, clapping fast, barely holding in excited giggles. She keeps at it—squat down, jump up and clap, squat down—moving her body in a dance between the human and the apian, earthbound and airborne inclinations. The youngest ones, twins, squat at a safe distance from the white sheet that has been spread out around the skep that the humans have offered, along with sprinklings of sugar, in an invitation to the bees to stay. From the edges of the scene, the twins watch and learn what to do when honeybees are bent on leaving.

The lesson, in short, is to talk to them and treat them like kin.[6] Kada appeals to the bees' mother because she knows that the bees will stay if their mother stays. The spell often works like magic, and the bees land nearby. Only then can a swarm catching proceed properly.

Professional beekeepers or bee biologists might point out that the swarm tends to alight low, soon after exiting, but this explanation should not discount the whole work of the binding spell or its success. The point is not just to catch the bees but to recreate a bond between the honeybees and humans at the very moment they are almost lost to each other.

The spell does that by chanting, invoking, and replaying the grounds of apian-human connection. It asserts their affinities: both bees and humans are embodied beings, dwelling on the earth between the soil and the skies, living in a language culture, and thriving within families and societies. The work of the spell affiliates the bees and humans into a cross-cultural, interspecies kinship as the beekeeper Kada addresses the bee mother, matica, as her own mother. Kada's children invoke the bee queen as their mother's mother.

The swarm spell is also an attempt to translate between species. The rhyme is meant to be compelling to the bees: the human language becomes more concrete—sugar, water shower, soil; more rhythmic—clap-pause-clap; and more elemental—soil, water, air. But it is not entirely wordless. The speakers presume that the local language, BSC, essentially expresses the apian *poots* and *kuts*.

Fig. 1.1 Sit down, mother

The swarm spell also simulates imminent threats: the tossed-up earth is a code for an earthquake or, as some locals have suggested to me, a plundering bear attack. The hose shower is the language of rain. Sprinkled sugar speaks for care in the form of supplemental food. The spell is partly a trick and partly an honest warning that upheavals, bad weather, and dearth await. The spell seeks to charm the bees into sticking with the humans with whom they live and work.

## *Local Styles of Beekeeping*

In local terms, Kada's beekeeping is known as *traditional*. Whereas most local beekeepers pride themselves on an avid reading of international apicultural sources—published in Slavic and European languages or in translation—and keeping up with the latest bee research, apicultural technologies, and worldly news, Kada never reads anything about bees. She draws instead on an oral family tradition and, when in doubt, seeks advice from other practitioners.

The most traditional aspect of her practice is that she lets the bees swarm at will and then chases after them every summer. She keeps bees in Langstroth hives, the familiar bee boxes seen around the globe, but unlike modern apiculturists, she does not use movable frames—which is the key feature of such modern hives—to manage the living order of the resident bee population, including their seasonal swarming inclinations.

Swarming, in the simplest terms, is an event of a hive's reorganization and apian social reproduction, a means by which bees rework their original community by splitting it up around new and old bee queens, or "bee mothers." The group that departs with the older queen ingests essential honey supplies before leaving behind the insulated, climate-controlled, safe nest that has become congested, contaminated, or otherwise uncomfortable in search of a new dwelling and a fresh start. The collective left behind may cast a few more swarms before it settles around a newly hatched queen. That queen will mate with up to a dozen drones, and from the sperm that she collects, she will be laying eggs for the rest of her reproductive life.

With their offspring, the hive will have an entirely different genetic composition than the original collective, and their social tendencies—toward foraging, swarming, cleaning, and such—will be significantly altered. The hive left behind is a hive transformed.[7]

As for the departing swarm, there is no guarantee that it will make it safely through this rite of passage that forms the search for a new nest. In fact, most swarms fail. Local beekeepers report ever fewer sightings of feral hives in the regional forests and claim that honeybees can no longer survive in the wild due to pest infestation, recurrent droughts, and prolonged dearth.

While the sight of a swarm, as one local apiarist put it, makes a "beekeeper's heart skip," most professionals cannot spare the time to negotiate with absconding bees, nor are they willing to run the risk of losing honeybees. Professional manuals soberly describe swarming as "wasteful," insofar as the swarming bees are allegedly distracted from foraging for honey and tend to cast too many swarms, most of which are too weak to provide for themselves in time for the winter. Still, whether to let the bees swarm or not is a highly contested issue among local beekeepers, who form a passionately opinionated community. Various styles of swarm management are tried out and advocated: from tight control of the bees' seasonal growth and brood-food arrangements that aim to curb swarming entirely to various compromises that work with the bees' schedules, priorities, and investments in rearing new bee queens. Nonetheless, the event of bees bursting out of hives is the clearest sign of "bees out of control."

By the looks of it, a swarm is indeed a riotous event; the sheer energy of restless, darting, quivering beings is contagious. Bees from neighboring hives join in, swept by the swarming currents, and human onlookers find themselves genuinely amazed. A word for swarming in BSC is *rojenje*, which invokes quickening and fizzling before coming together. One says, for instance, *misli mi se roje*, "My thoughts are swarming," in a rush of an inspiration or in an anxious flurry.

Swarming, however, is just as much an act of bees' deliberation, long in the making, timed according to their sense of opportunities and priorities. Preparations for the swarm's departure, as bee biologists and beekeepers know, are very complex and still inadequately understood, and the communication, coordination, scouting, and decision-making that follow are just as ingenious and nothing short of fabulous.[8]

A swarm is the incipient form, and a recurrent beginning, of an exceptionally sophisticated, deliberate form of a well-knit togetherness, which is the hive. The political form of the hive—political in the sense of being together through experience and language and working out a communal life—in the local language is not a "colony" but a *društvo* or a *zajednica*, the terms with thick regional histories that translate to "society" and "community."

Along with these Slavic terms, the idea of *umet* also defines bees locally. Like many other key Islamic terms among Bosnian Muslims, *umet* is borrowed from Arabic (*ummah*) and denotes a faith-based community.

My interlocutors with theological training referred to a particular verse in the Qur'an to underline commonality between humans and bees. Found in the chapter titled "The Livestock" (*Al-A'nām*), the verse refers to animals in general and translates as follows: *All the animals walking upon the earth and all the birds flying on their wings are communities like yourselves.*[9] Classic Qur'anic explanations of this verse offer a wide range of interpretations that tend to meet on several basic points: animals are endowed with their own languages and engage in forms of adoration specific to their species. While all animals are an *umet*, the intricacy of honeybees' social organization, the complexity of their hive products, and their mention in the Qur'an are the usual arguments that apiarists put forth to single out the honeybees as an especially faithful and knowledgeable community. "Honeybees know best," the local beekeepers are fond of saying.

## A Prophetic Species

The sense that bees "know best" when it comes to running their hive affairs is commonly shared across many styles of local apiculture. The extent to which human control over bees is curbed by this underlying confidence in the bees, however, varies. Kada's hive management is minimal, but her spell pleads with the bees because when a swarm is cast, human control is, obviously, not an option. Even once a swarm is caught, there is no guarantee that the bees will stay. With a casual invocation of God's name over the newly established hive, Kada acknowledges what seems like common sense to her: the fact that bees are, first of all, divine subjects.

She appeals to the bees to settle, but she appeals to the God that she and the bees have in common, the Compassionate and the Merciful, to bind them together back at the apiary. For Muslims, the invocation of God's names is a preface or a coda to any act of significance. When followed by *alḥamdulillāh*, "Praise be to God," and *rabbi l'ālamīn*, "Lord [or Nurturer] of worlds," a casual supplication shows good manners and states faith in good outcomes soon to follow. For out of kindness, the tradition guarantees, God will respond grandly to such praise. The two lines of invocation and praise make up the quintessential Muslim address. They begin the chapter that opens the Qur'an and are recited in daily ritual prayers and everyday supplications. Kada, who is not much drawn to perusing the Islamic sources—she goes by the handful verses she learned by heart—takes it for granted that honeybees work and live by means of divine inspiration as well as that they associate with humans only for the love of God.

These are the basics of an Islamic understanding of bees, which profoundly orients local beekeeping, whether through the lively oral culture or the rich textual tradition. Among the beekeepers I have worked with over the years were many working and retired imams with a formal training in what is known as "Islamic sciences," which include diverse subjects of study such as Islamic theology, Sufism and philosophy, ethics, and jurisprudence. A few beekeepers were Sufis: Muslims drawn to the inner path to knowing and loving God. For the Sufis and imams, the canonical sources—the Qur'an and the Hadith (records of the Prophet's words and deeds)—and their commentaries as well as the classic and contemporary Islamic texts are a part of the written tradition that is just as pertinent to keeping the bees as are the beekeeping manuals.

The Qur'an describes the event of honeybees receiving a divine revelation. Two verses in the Qur'anic chapter titled *al-Naḥl*, "The Bee," are commonly cited and recited in learned conversations about honeybees. In translation, they read: *And your Lord revealed to the bee, 'Take up dwellings among the mountains and the trees and among that which they build. Then eat of every kind of fruit and follow the ways of Your Nurturer made easy.' A drink of various hues comes forth from their bellies wherein there is medicine for humankind. Truly in that is a sign for a people who reflect.*[10]

The divine message revealed to the bee, at the very least, amounts to guiding instructions on how to forage and where to dwell. Foraging entails biodiverse landscapes—*every kind of fruit*—while divine bidding makes residence in apiaries incumbent on honeybees.

The bestowal of revelation on the bee is most noteworthy. Whereas divine guidance, the Qur'an says, is granted to every engendered thing,[11] the word *waḥy* and the verb form *ewḥa* from the same root in the Book primarily denote divine revelation or inspiration to human prophets and close friends. The prophetic species, honeybees follow the revelation unfailingly—in contrast to humans, my interlocutors often added regretfully, who can err or defy God's words—and their inspired ways of living yield the cherished honey.

Given honey's mention in the Qur'an and the pride of place it takes in what is known as the prophetic medicine—health advice passed on by the Prophet of Islam—the therapeutic and nutritional properties of honey have long been acknowledged in the regional herbal folklore and remedial home diets. Collection of honey has traditionally been the primary focus of the local beekeepers. In addition, since the 1970s, therapeutic properties of the whole array of other hive products have been appraised in a transnational effort to ground traditional uses of bee products in scientific trials, explanations, and proofs of efficacy.[12]

Consumed alone or combined with herbs, hive products, with honey at their core, are a part of nutritional, prophylactic, and medical regimes that people in present-day BiH adopt in case of illness or to aid a recovery. At the very least, a regular supply of honey is what most people aspire to secure for the sake of a sound diet.

The medicinal qualities of honey, however, have a still broader connotation in the Islamic sources and their local commentaries.

The Arabic term *shifā'* (a cure or medicine) is used in the Qur'an to refer to honey and to the Qur'an itself, both fruits of divine revelation. The drink that comes from the bees' bellies is a shifā' (a healing), while the Qur'an is a *guidance and healing for the faithful* and a *cure for that which lies within breasts*.[13] As a verb form, *yashfi* appears twice in the Qur'an to relate God's healing touch. The bestowals of divine revelation and their remedial qualities in honey and in the Book that is meant to be read and recited in a beautiful—sweet—voice are staples for reflection.

The Qur'an urges its readers and listeners to ponder the meaning of the verses and to take a good look at the signs of the material world.

Fig. 1.2 And Your Lord revealed . . .

For Bosnian Muslim beekeepers and bee lovers—imams and Sufis in particular—such reflection is a part of faith and of the apicultural practice. A nimble habit that travels from apiaries to home libraries, reflection ongoingly refines a growing yet unfinished understanding of bees through both hands-on observation and the contemplation of meanings across the pages of the Revelation.

This back-and-forth between the world and the divine Word is presumed by the Islamic devotional and reading tradition and suggested by the double meaning of the word *āyah*, which glosses both a thing out there in the sensuous world and a verse in the Qur'an. God reveals His presence everywhere and speaks through every form.[14]

## *The End Times*

In the final days of the final times, the Qur'an will vanish from Earth overnight, one hadith reads.[15] We are left to imagine the details. Every letter on every page will rise up on cue and flee. Every electronic transcription will abscond from the code stored online or on hard drives. Like bees who receive a message across the nest, no one knows from where or how, then surge one by one, together as one in a swarm to leave congested or infected hives. Like a giant swarm, the words of God will depart to the heavens for good. "No verse will be left on earth."[16]

The Prophet's saying speaks of the world on the brink of collapse, but the world that is essentially already ruined because the revealed knowledge has been practically forgotten and the Qur'an has already disappeared from people's hearts. Revelation is revoked when human desire for divine news has been spent.

A body of the Prophet's sayings speaks of social, political, and environmental upheavals that signify the coming end of the world. The final times, however, will be a long-lasting epoch. The Arabic phrase *ākhir al-zamān* (transliterated in BSC as *ahiri zeman*) refers to the historical period that began in the seventh century when the Prophet of Islam delivered God's final word: the Qur'an. The last Prophet and the last revelation announce the beginning of the end of Earth's history.

The Hadith speak of major and minor signs of the times, but their iterations across the chains of transmission can hardly be used to chart the exact course of historical events. That has not prevented Muslims over time from trying to work out a chronology of collapse. At their most ambitious, such attempts seek to derive a definitive sequence of affairs and a conclusive cast of characters from the materials whose suggestive power rests in their multivalent nature as well as in the gripping details that impart ultimate significance to everyday disasters.

The signs of the times described in the Hadith, however, do suggest a worldwide dynamic of a race to ruin. Capital expands, consumption is exuberant, fantastic technologies manage the sky and the land, and eminent swindlers paint the hell of a present as heaven to win a global fan base. There is much useless killing. Punishing droughts persist over wastelands, and land "sinks" in the East, in the West, and on the Arabian Peninsula. Thick smoke envelops the atmosphere. When profound disorder fills the earth, the cosmos heaves in response and revolts once and for all. The sun rises from the West.

As with all other matters, the Prophet's sayings on the final times are meant to be taken as an extended commentary on the Qur'an whose ample references to the End, convey a bottom line: it is God's well-kept secret. The Hour will come, God promises, when least expected: *in a blink of an eye, or faster still. Indeed, God is able to do all things.*[17] Not even the nearest angels know the particulars. They can do no better than watch for the signs and dread the event that *weighs heavily on everyone in the heavens and the earth.*[18] The same is true for the Prophet, God's closest friend. *They ask you about the Hour: When will it come? You don't know, so how can you tell. To your Lord alone belongs the knowledge of it. You are but a warner for whoever fears it.*[19]

What the Prophet conveys instead is how to relate to the grave divine promise, day to day. The answer, in a nutshell, is with keen mindfulness. Whenever the Prophet raised a morsel of food to his blessed mouth, a hadith conveys, he feared that the End might surprise him before he swallowed. And he was just as expectant of his death. If there was no water to perform ablutions while traveling and the nearest well was an hour's journey away, the Prophet performed ritual purifications by touching soil, for fear he might not live long enough to reach the water source.

My principal guide through the subjects of Islamic thought and practice was a wise, gentle Sufi elder, Shaykh Ayne. For him, the final times was a default reference to our present, while eschatology was the cornerstone of our conversations, not as a discrete subject but as an underlying awareness about the finite nature of all things and a weariness about the coming Hour.

Shaykh walked me through the seminal texts—the Qur'an and the Hadith—and taught me to read them for their timeless messages and insights into the present. He was uninterested in teasing out the unfolding of the apocalyptic plot but was deeply intrigued by the human dispositions and collective drives that obstinately wrecked and ruined the earth at present to the point where the Gentle and the Patient, the names by which God is known in the tradition, winds up the planet, the cosmos, and all the engendered worlds.

One book in Shaykh's library particularly impressed me. Entitled *The Path of Muhammad*,[20] the book is a work by by one of Shaykh's teacher's, Mustafa Čolić, an imam and a Sufi of an extraordinary insight. Published in 1998, nearly four hundred pages long, the book is a commentary and a translation of an excerpt from the sixteenth-century classic by Sufi scholar Imam Birgivi.

Shaykh Čolić's book reportedly sold like hot cakes, though, as Shaykh Ayne amusedly noted, the general agreement was that Shaykh's comments were too challenging for most readers, including those well read in the Islamic sciences. In his role as an imam, a knower of great repute, and a witty speaker, however, Shaykh Čolić regularly lectured and conversed in public and private gatherings, and so his speeches took this dense text onto a well-traveled life course. As so often happens in Bosnia, Islamic textual knowledge flows off the pages, trimmed but also enlivened and clarified by the spoken word.

The section dealing with the signs of the times, a mere fifty pages, sits among discussions of a range of other Islamic creeds derived from the Qur'an and the Prophet's sayings. All the signs of the times, Shaykh suggests, can be taken both literally and as allusions to deeper meanings, depending on the listener. The readers' particular positions and personal dispositions color the meanings, like glass colors the water or light it receives.[21]

Appreciation of multiple perspectives, however, does not mean that anything goes. A sound understanding of Islamic sacred law and the creeds is the common ground on which to build further understanding. Dogmatic interpretations are futile, Shaykh writes, because they strive against the evident plurality of perspectives and deny the multiple nature of the world itself, which comprises both sensuous things and the imperceptible realities that compose and surpass them. Reductive, or dogmatic, tendencies are raging, according to Shaykh Čolić, and include rationalist and formalist currents of modern Islamic thought, which condemn a pursuit of metaphysical meanings as well as socialist ideologies of historical materialism, secular Western philosophies, and "godless sciences and technologies," which deny all but biophysical reality confirmed by human senses or instruments.

The whiff of the great smoke that the Qur'an mentions arising to envelop the earth, Shaykh writes, is already here, miasmic in the industrial air pollution. The Arabic word *dukhān*, translated as fume, smoke, gas, or dust, can be taken literally as that which pollutes the air and obscures the vision in the wake of fires and storms. Early commentaries suggested that *dukhān* referred to the great famine that struck a tribe of the Prophet's opponents, clouding their vision. The greatest fumes and fogs at present, Shaykh writes, are those occluding minds and deadening hearts with knowledge that tries to reign in the expansive range of ongoing divine self-disclosure to the measure of what humans deem possible and plausible.

Shaykh Čolić's commentary on the signs of the end times gives valuable and timely insight into the nature of the disorder that ultimately dooms the planet. "Since all that is manifest points to what is non-manifest, all objective events and things at once point to subjective activities and matters, therefore, all material catastrophes allude and stoke contemplation towards metaphysical (related to mind-reason, spirit-soul) catastrophes."[22]

The manifest and nonmanifest are complementary qualities of the reality that both integrates and surpasses them, and their dynamic interplay is an ongoing subject in Shaykh's writing on Islamic metaphysics. The point I take for this book, however, pertains to the nonevident catastrophe at the core of final times, which has to do with human dispositions. Human (*insān*), in the sense that Shaykh elaborates through his works, is a being special in the light of the inner domain glossed in the Islamic sources as "heart."[23] Heart is, potentially, a tremendously capacious entity where the inner and outer, the manifest and nonmanifest, God and human meet. The human heart can also fester with disorder that incurs more damage, on the inside and the outside, "than seventy devils together."[24]

Shaykh's comments also suggest that the portents of *ākhir al-zamān* are always timely, abounding in particular historical connotations and foreshadowing the ultimate events at the end of time. Signs of the times, in Shaykh's reading, are the signs brewing at all times.

## *Eco-Eschatology*

What follows is not a comprehensive analysis of the Islamic teachings on the final times but rather an attempt to take on the atmosphere of the looming End to ponder the timely signs of the unfolding ecological disaster. I take a page from Shaykh Čolić's book to think about our dire present with honeybees and their keepers, their faithful, seasonal pursuits of honey and bees' famished foraging through weathered landscapes. I also study Islamic sources with local fellow readers, listen to the popular as well as learned commentaries, and, just as importantly, draw on years of field observations gathered into the beekeepers' and my notes.

Focus on honeybees helps me open up a special window onto the Islamic eschatology, one that views the final times ecologically through relationships between humans and animals, among other species, and God.[25] An eco-eschatological perspective attends to Islamic textual sources to better understand Bosnian Muslim tales and concerns about the bees and the changing planet. At the same time, it is the particular Muslim eco-practices and commentaries on the field evidence of climate change crisis that highlight elements of the Islamic tradition that are often underappreciated in contemporary conversations about Islam.

I also read across humanities, social, and natural sciences for critical insights into our catastrophic times. The worldwide ecological and climate catastrophe and the evident failure to effectively curb global emissions and warming trends have prompted academics and public intellectuals to scrutinize our anthropocentric—human-focused— ways of knowing and relating to nature. Within humanities and social sciences, self-reflective scholars have lately been keen to shift the focus of their studies beyond the modern idea of the human as a sovereign creature entitled to use with impunity everything between the land and the skies. Current works in interdisciplinary scholarship and the visual arts are now highlighting the myriad ways in which animals, plants, and other species as well as natural elements deeply shape our histories and economies and our human ways of being, knowing, and dwelling within the world shared by many species.[26]

My Islamic sources and Muslim interlocutors, however, will not let me lose sight of divinity, which is seldom of concern to the burgeoning field of multispecies ethnography and human-animal studies. The divine is not simply omitted, but it is ill fitting within the field of the new eco-scholarship. Namely, scholars and artists are becoming conceptually adventurous because moving beyond the modern idea of human entails a readiness to duly consider other propositions about the human place in the world. Indigenous and nonmodern cosmologies that describe landscapes as imbued or animated by spirits and nonhuman beings are now given another chance at being heard. Truth claims that cultural relativism could once accommodate and, for all practical purposes, dismiss as irrelevant are now recognized for their potential to disturb modern certainties about what exists and how things are, as well as to effectively counter land-use policies, extractive enterprises, or conservation plans that rest on the assumption that the environment is merely an economic resource.[27]

Monotheistic cosmologies, on the other hand, tend to be dismissed or glossed with hastiness untypical for scholars.[28] The idea of One God in the Abrahamic traditions, not as the subject of a personal, religious belief but as an expansive agency to reckon with, cannot be shoehorned into the burgeoning visions of the plural world, the multiverse in which many forces and claims about what matters can coexist, but not one can get the last word. For all its expansiveness, this "new age of curiosity" draws the line at the subject of revealed religion, which arguably remains the most discomfiting, indeed, irritating, issue for secular thinkers.[29]

Similarly, metaphysical questions—about ultimate values and the meaning of human, presence, or the world—are being widely reopened in the light of ecological catastrophes and the prospects of postsustainable future. The metaphysics that now come to the fore, however, are the ones newly invented and vocally opposed to the old, theological traditions, that, assumption goes, are conceptually threadbare and, anyhow, nearly transcended with globalization of modern secularism.[30]

Rare attempts to think ecologically with the Abrahamic religions often proceed by reading that is openly unbound by the tradition, and tones down most inconvenient elements, such as the apocalypse or the idea of a resourceful, competent God whose knowledge encompasses the future and the fate of the earth and the universe and their dwellers. Ecological rereadings of Christian theological sources in particular tend to present a minimalist metaphysics as the very condition for considering—generously and seriously—religious texts, experiences, and utterances.[31]

This book intends to reread Islamic sources, reverently, treating them as founts of theoretical inspiration. In other words, I think with the conceptual tradition that is obliging Muslims—and their anthropologist—with metaphysics in its most expansive sense, including the idea of God who promises to bring the world's end. I suggest that ecological thinking and antiapocalyptic activism can arise precisely from an engagement faithful to the boldest claims of the Revelation, and that religious traditions can nurture adventurous thinking among many people, not just professional scholars.[32]

This book proceeds by setting forth two related popular eco-eschatological propositions. One is that honey is waning, which is both an empirical statement about the effects of climate change on local landscapes and a weary comment on the profound meaning of the planetary crisis. The second proposition is a that honeybees' endangerment foreshadows the world's collapse, a confident assumption among Bosnian Muslims that does not merely rest on the adverse effects of honey's scarcity on hives' health. What is it about honey and bees that makes them vital to the world? Why are they invested with such an ominous capacity?

My point of departure for entertaining these questions is the hadith about the vanishing of the Qur'an. The Prophet's saying forecasts a bleak moment when God's revelation finally withdraws from Earth—the sign that the wasted planet is doomed. This saying is one of many that concerns the state of knowledge in the final times and anticipates a historic moment when publications are prolific and a mass of information is circulating, but genuine knowledge is on the wane. In Shaykh Čolić's commentary, the knowledge in decline refers to several related forms of grasping the meanings that are not obvious. To begin with, there is *ilum*, the knowledge of Islamic sciences; *tesawuf,* taken broadly as a nonreductive interpretation of religious teachings, and *marifet,* which refers to deeper forms of insight found through contemplation and inspiration on the path of seeking God.[33] What makes these forms of knowledge "genuine" is the extent to which they are true to the Revelation and keep it a lively resource. A resource for grooming the hearts, which receive news of God and are seats of knowledge about God. An eco-eschatological reading of this hadith about the vanishing of the Quran and its commentaries proceeds from the premise that within the Islamic worldview, divine revelation has always been more than the written word.

There is also the divine message bestowed on the bees, the fruit of which is honey: *the healing and the sign for people who reflect.* For there to be honey, however, the entire world is presumed to be coming together—skies with soils, plants and proboscises, nectar secretions with seasons—to nurture the prophetic species.

Entities and actions within soil—minerals, acids, nutrients, and resident rock types—draw particular plants and prime their nectar-secreting tendencies. Local clouds and wind currents are mercurial elements that spell out density, sugariness, and chances of nectar flow. Foragers track the sun's course across the sky to map round-trip routes from nests to nectar sites, with an innate solar compass that connects hives to the center of the solar system even when the sun is obscured or invisible. Earth's magnetic fields resonate with bees in flight, perhaps by means of a magnetic element in their abdomens (scientists are debating), but whatever the structure of apian magnetorception, there is more than a poetic resonance between the micro and the macro, the bees' bellies or bodies and the earth's innards where geomagnetic fields spawn and fizzle before extending into outer space to fend off "solar winds."

This coming together is the way the cosmos—that rattlebag of everything other than God, from weeds to solar winds—responds to tireless divine bidding and displays godly presence: *All those in the heavens and the earth ask of Him; every day, He is upon some task.*[34] Islamic thinkers usually speak of the cosmos as *al-kathra*, the evident plenitude of differences, all of which are revealing—and veiling—the One unique divine reality that engenders them and ultimately integrates and transcends them. The Qur'an presents God as the Lord of the Worlds and points to scriptures and to the things in the world as manifest signs of God's actions and attributes. Shaykh Čolić speaks of divine self-disclosure in complementary forms of the worlds and the books (*alemi, kitabi*), or of beings and words (*ekvani, eklami*).[35] In other words, the world is the sensuous news that God delivers at all times and a close regard of these signs is a counterpart to listening and reading the Qur'an.

The waning of honey, divine revelation at its sweetest, is analogous to the vanishing of the Qur'an from Earth. For the premonition of the hadith to make ecological sense, one needs to wonder, what is divine revelation to the world? What, in this worldview, could bring down the world?

The rest of the book fleshes out an eco-eschatology, showing how ecology and eschatology are related in the Bosnian Muslim every day, how modern ecological thought and Islamic sources complement each other, not least in their divergences, and why ecology these days cannot afford to ignore eschatology, just as no religion of any contemporary relevance can ignore the global ecological disasters underway.

## Honey Ecologies

Pollination ecology refers to pollen flows within a habitat that mostly take place through plant-pollinator interactions. Key events in life cycles of plants and insects are timed to a host of environmental cues and well synchronized between pollination partners so that the flowers' pollen offerings readily meet adult insects' appetites.

Given the importance of pollination for reproduction of plants in both natural and agricultural ecosystems, pollination ecologists throughout the twenty-first century have been preoccupied with the growing anthropogenic assaults on plant-pollinator interactions, from habitat loss and pesticides to invasive species and insect declines. The implications of global climate change for pollination networks and their conservation are still inadequately researched.

Later onset of winter and earlier springs are among the most obvious indicators of global climate change, and how the coupled plants and pollinators respond to these trends is an open question. This prevailing uncertainty is partly due to the fact that experimental and field studies are still rather few, considering the number of species involved and the specificities of their relationships in different climates and environments. Difficulty of anticipating future trends also stems from the sheer variability of responses noted thus far, including both synchronous adjustments of plants and their pollinators and pollination partners' lives growing out of sync.[36] While pollination networks are generally described as robust, pollination ecologists suspect that the potential for plant-pollinator decoupling is great, especially with future warming and extreme weather trends.[37]

Local beekeepers' interest in seasonal interactions between bees and plants, on the other hand, is practical and focused on nectar. They also keep an eye on availability of pollen in local landscapes, since its collection by the bees is essential for the hive's nutrition, brood development, and, ultimately, apian survival. To improve the chances of honey flow, many beekeepers enthusiastically plant flowering trees and plants and collect and cast seeds of melliferous—honey-yielding—wild bloomers in the areas surrounding their apiaries.

Mobile beekeepers travel with their bees across the country, actively following seasonal blooming trends. Honey forecasts are sometimes gauged with electronic hive scales. The scales are fitted to trial hives to track nectar inflows in the field and send updates to the apiarists' cell phones. Most reliable anticipations, however, come through beekeepers' scouting of familiar or new areas. Cruising and on foot, they closely watch the plants' development and the ambient weather for the signs and prospects of honey flow.

When I started fieldwork in 2014, a local saying that "honey's on the wheels" summed up an agile strategy to chase after honey that could no longer be counted on to flow where the apiarists' hives were poised to meet it. Not even avid planting could secure honey harvests. Worse still, forage yielded so little nectar that the bees continually needed supplemental feeding. By 2019, honey was becoming so scarce that even those who traveled in pursuit of it were recurrently disappointed.

The rest of the book describes how beekeepers cultivate conditions for flow and hunt after honey as well as how they handle dearth and disappointments. Chapters conveys beekeepers' savvy observations within bee yards and across countrywide forage sites, learned from many years of experience and informed by deep knowledge of the insects and their favorite plants. Apiculturists' interests and notes paint a picture of honey ecologies, which are nectar-minded but are also, by default, mindful of wider conditions for nectar secretion and the bees' ability to forage.

As such, honey ecologies add complementary concerns to the vital and still emergent inquiry into the effects of climate change on living landscapes. Local beekeepers' concerns point beyond the timings and durations of the seasons—namely, earlier offset of spring, extended summers, and the later winter—to chronicle instead a hurried trend of disturbance in the quality of seasons. Beekeepers' observations of honey ecologies in the weird new weather suggests that vital relationships between elements, plants, and insects are not just mismatched but tattered in many different ways.

To begin with, seasonal outlooks are becoming hard to predict. Spring tends to come earlier every year, but it can also be significantly delayed. Winters are interspersed with "false springs," the events that send premature warming cues to bees and plants and are often followed by a drop in temperature, frost, or snow. Spring no longer appears to be a season that arrives with definitive atmospheric and environmental changes but rather is a mixed-up season, sometimes jump-started with snowmelt and sudden spikes in temperatures, then stalled by spells of cold weather. Air temperature in the spring tends to be unseasonably high—high enough for the nectar of spring bloomers to evaporate—while rainfall patterns dramatically vary from year to year. Past summer droughts extend well into the spring, or cold, extreme rainfall lasts into the summer months. Season as a category now barely holds. As one seasoned beekeeper, Sead, puts it: "Spring is no longer spring, summer is no longer summer, winter is no winter. What we have now are extremes."

When we met in the summer of 2016, Sead was a retired electrical engineer with a well-kept beekeeping diary. He was comparing the recent years at his small apiary in western Bosnia with what beekeeping used to be over his forty years of experience with bees. "I swear to God," he says, "something arrived in nature that changed everything. Everything." Unseasonal and extreme weather changes interfere with bees' foraging and plants' flowering and nectar secretion in various ways, some of which are obvious: flowers suffer from frostbite or heat damage, pollen and nectar offerings are washed off, and bees are rained in.

Next, apiarists have noticed that the duration of bloom and nectar secretion is much shorter than it used to be. Mehmed, a resourceful beekeeper from central Bosnia, remembers that the ringlets of snow-white black locust—the first strong flow in the region that yields thick nectar with a glow of rose gold—used to last three weeks: "One week the florets are opening, next week they're in full bloom, and the third week they're waning." Now, "within a single week"—he claps his hands at this—"they've come and gone."

Furthermore, beekeepers note a greater variability in weather and forage conditions in landscapes within relatively short distances. Traditionally, mobile beekeepers exploited a rich range of microclimates across BiH that spelled out different forage possibilities. The black locust, for instance, is known to flow early and copiously in the riverine climate of the country's northeastern regions. With the plant development and nectar secretion becoming increasingly uneven, nectar prospects now vary greatly within the same forage area, and beekeepers are pressed to seek out nooks that seem particular auspicious and quickly move with the bees should nectar fail to flow on a given site.

What is more, blooming schedules of local species of trees seem scrambled. Many beekeepers cultivate different varieties of fruit trees to extend the duration of nectar flow within a season. Mehmed, too, has lovingly planted the bee yard by his home with apple, pear, and cherry trees. Among them are early, midseason, and later bloomers, but lately, he reports, all varieties can bloom at the same time. Late bloomers in the orchard can also surprise him with the earliest blossoms.

Differences in blooming schedules for the same species of plants across microclimates and at different altitudes have always been important to beekeepers. Mehmed used to wait for the blossoms of the South European flowering ash, known as manna ash, by his bee yard to wane before he transferred a batch of his hives to the nearby mountain. That way, his bees would arrive just in time to catch the beginning of feathery yellow mana ash bloom, customarily delayed at an altitude of roughly a thousand meters. Nowadays, he says, baffled, mana ash can bloom at the same time at different altitudes. Or the highland varieties can bloom first.

Overall, summer temperatures are generally extended into late fall, but contrary to expectations, the extended summers cannot be presumed to afford longer growing or foraging seasons. In the aftermath of heat waves and droughts in the region, autumnal landscapes are left withered. Nectar is scarce. Instead of being winter-weary and preparing for hibernation, hives continue active flight and brood building well into December. The hives' residents at this time, known as "winter bees" that are typically fewer in number but longer-living, are spending their bodily fat reserves and consuming hive winter food stocks. Consequently, food supplies may be low when the winter sets in and the clustering bees need them most, while the resident bees may be too exhausted, literally, to build the hive back up come spring.

And the strangest of all. Beekeepers are consistently noting days and weeks when environmental and atmospheric conditions seem perfectly conducive to flow and yet hives are empty of nectar. Or, bees are ignoring what seems like an ideal forage. Quoting Mehmed again: "There are chestnut trees in bloom, their blossom crests thick, like startled cats' puffed up tails, but you look and see there's not a single bee on them! Incomprehensible, what is going on? The blossom like that, worthy of every attention, the blossom to admire, and you think to yourself, 'It's impossible that it has not a drop of nectar, not a speck of pollen.' You look, and there's nothing. What is going on? That's the greatest puzzle."

Other local beekeepers also express their bafflement at the signs of changes they are noticing. Pulling on their expertise, they sometimes speculate about particular circumstances that led to a failure of forage, but they tend to frankly admit the limits of their knowledge, especially when the events are without precedent and the usual signs are off. Mehmed thinks it is up to scientists to further investigate the strange new trends, but he insists that the disturbances newly noted are not the beekeepers' problem: they ought to concern us all. "What is going on?" Time and again, he raises this open question.

### *Multispecies, Muslim Hope*

"*Ākhir al-zamān*?" I ask Mehmed. "Of course it is *ākhir al-zamān*," Mehmed replies, then makes this matter-of-fact statement open ended: "But still..."

What I learned in the course of my study with the local beekeepers is that climate change and the final times are both relevant descriptions of our present, but each only marks the beginning of an inquiry. "What does it mean when plants no longer attract insects, that there is no nectar? Do the plants lack desire, or do they lack conditions for reproduction? What is going on? And the main question: Is this telling us that the majority of plants will disappear?"

Questions such as Mehmed's that arise from local honey ecologies undergoing sweeping change are important because they state the many great unknowns entailed by the rapid alterations that are churning the biosphere. The greatest insights that emerge from current scientific studies of climate, ecology are, likewise, not definitive answers or confident projections of the future trends but rather thoughtful brainstorms on the multitude of elements that beg consideration, honest reflections on the gaps and blind spots in the conventional studies, methodologies and theories.

Mehmed is devoted to the honeybees and has been a full-time beekeeper ever since he withdrew from his post as an imam in the local mosque. His university training in Islamic sciences, he says, brought him closer to the bees. Over the years of our friendship, I have observed the ways bees keep up his interest in the Islamic sources. Mehmed and I discussed Qur'anic exegesis in his bee yard and poured over books on the veranda of his home overlooking the hives. But more tacitly, keeping bees was another of way of keeping the faith alive within an eco-practice that referenced Islam as a wordbook of values and inspirations to counter disasters and disarm despair.

"Honeybees are still doing what they are inspired to do. They've accepted the revelation and are working accordingly. She [the pronoun for bee is "she" in BSC] keeps searching for nectar, even if there isn't any. I watch her, she too is hoping. She too is hoping, I see it, because she keeps trying hard."

Mehmed interprets the honeybees' striving as "hope." His interpretation presumes both that there are core similarities in species relationship to the world and that some conducts and dispositions—such as striving and hope—are properly Islamic across the species of divine subjects. Qur'an describes animals and plants—but also things that modern people consider inanimate, such as planets, wind, or stones—as divine subjects who live in Islam, that is, abide by divine commands and praise their Lord. Mehmed takes this multispecies Islam for granted then goes further by noting that bees, too, are hoping, Hopefulness is explicitly recommended as a part of faith—to humans, for humans alone can succumb to despair. To despair is to doubt divine mercy, which the tradition describes as exceedingly vast and encompasses all things.[38]

There is no giving up, Mehmed says at the tail end of 2021, another dismal year for nectar flow. He describes the preparations he undertook for the next year—he helped new swarms build up food stocks for the winter, among other things—and the forage sites he planned to visit again. Honeybees, Muslim beekeepers usually say, are an admirable source of inspiration for humans, but on this occasion, Mehmed does not explicitly say that he takes cues from the bees. Rather, he describes the perseverance that he and the bees have in common, though nectar was a mere trickle for several years in a row. Under the dire circumstances, the thing to do across the species is to strive faithfully.

Fig. 1.3 Spring is no longer

## *Quiets of Disaster*

"Have you heard the joke about God deciding to bring about Kijamet [the End]?" a taxi driver asks my sister and I. Curious, we shake our heads at his gaze in the rearview mirror. "So," the taxi driver proceeds, looking pleased at the chance to tell the joke, "one day, the time has finally come and God glances at the world. The first thing he happens to see out there is Bosnia. 'Hmm,' God says, 'it looks like I've already been at it, there.'"

The joke bespeaks irreverent ways in which some contemporary citizens of BiH invoke God and apocalypse. BiH is a multiethnic country where three dominant nationalities—Bosnian Muslims, Bosnian Croats, and Bosnian Serbs—are formally associated with Islam, Catholicism, and Eastern Orthodoxy, respectively. Secularism is also an enduring legacy of Yugoslav Socialism, and religious lives and institutions have been both revived and, some suggest, compromised by their often overt associations with political parties and issues of governance.

Within the Croat-Muslim Federation, one of two entities of BiH in which I focused my study, Bosnian Muslims form an ethnic majority, but Muslims in Bosnia, as anywhere, display various forms and degrees of practical commitment to Islam. In other words, being Muslim does not imply a default familiarity with the textual tradition or adherence to religious rituals, let alone an interest in cultivating an Islamic *ahlāq*: a virtuous character. Among the people I worked with are imams and Sufis who tend to be exceptionally devout and conversant with the Islamic tradition as well as Muslims for whom religious teachings are conveyed through a largely oral culture and who abide by the mores and rites of the tradition more selectively. Or not much at all.

General statements about a Bosnian way of being a Muslim are therefore bound to be imperfect. In the following pages I have particular people in mind and will be citing their words and describing their practices, but I am also interested in making broader claims about Islamic ecological or eschatological insights in Bosnia and beyond. In doing so, I read texts and listen to the tales for what the tradition recommends in a way of living a life. I also hope to capture a whole range of lived orientations toward the Islamic tradition that thoroughly saturates Bosnian culture, even on occasions when it seems least obvious or in instances that seem particularly irreverent.

The keyword in the taxi driver's joke is Kijamet, a local transliteration of the Qur'anic Arabic word *al-qiyāmah*, one of many names that designate the eschatological events of the world's end. *Al-qiyāmah*, in particular, refers to the standing after resurrection in hopeful or fraught expectations of divine judgment. Colloquially, Kijamet in BiH glosses the apocalyptic event itself, as in the event expected by the two angels in the tale begins this chapter. Kijamet also locally refers to catastrophes of any sort and to catastrophic weather events in particular.

Invoking Kijamet as a mundane catastrophe, the dark joke can be heard as a bitter commentary on the quotidian reality of a country that has not recovered from the 1990s genocidal war and the collapse of socialism. Many of its citizens feel bogged down in the political stalemate, ethno-national warmongering, economic precarity, and growing class disparity that prevail since the internationally brokered peace agreement. The population of the mobile—young, educated professionals and blue-collar workers—is massively emigrating in search of jobs and better lives. I hear in the joke a reference to the state of a day-to-day crises, a life lived through low-key emergencies, the quiets of disaster, and disorders that are chafing and undermining existence in the wake of loud catastrophes and their aftereffects from the war to the storms.

Like all good jokes, the taxi driver's is a capacious comment that leaves room for multiple interpretations. When attention shifts from the country's political situation, which tends to preoccupy the media and the scholars of the region, the disaster the joke denotes is environmental.

BiH is heavily dependent on coal for electricity and winter heating and, at the time of my research and writing, was expanding the national fossil fuels energy sector. While the country's contributions to the global $CO_2$ emissions are comparatively negligible, the outdated power plants are routinely polluting local air in gross violation of global environmental standards on emissions of pollutants while also generating deposits of toxic sludge and mining waste that are likewise sources of serious public health hazard. Coal and far dirtier substances like heavy fuel oil or automobile tires also heat businesses and households through the winter.

The country is often described as "postindustrial," although cities around BiH are currently suffering heavy pollution from privatized units of former socialist enterprises. Compared to the socialist era, the dirty industry at work is now small in scale, but the ongoing production—of industrial coke, steel, cement, cellulose, and leather, to mention a few sectors—is environmentally unregulated while its waste is disposed of haphazardly: to compounds, rivers, forests, and to third parties without technologies of waste management and without liabilities.

In addition, former industrial zones across the country are laden with highly toxic chemicals, such as lead, chlorine, polychlorinated biphenyl, and by-products of toluene diisocyanate (TDI), to name a few. These lethal remnants of socialist enterprises are stored in rusting, leaky containers and hastily buried or deposited in the open air. Exposed to weather, contaminants move with soil, wind, and rain. Absorbed by plant roots, leaves, and nectaries, toxic loads imbue pollen dust and flow with nectar through the honey seasons.

Due to disastrous environmental policies and practices, and considering that the national administration of resources and infrastructure is diverted by the ethno-national electoral priorities, the country is exceptionally unprepared for extreme weather and for the battery of other adverse effects anticipated with the future of climate change.

*Forage Frontlines*

Local beekeepers try to avoid industrial and urban pollution and the rampant new peri-urban development and seek out heavens of "genuine nature" (*prava priroda*). They establish apiaries in mountains, in forests, and on village outskirts. Mobile beekeepers tour the country's back roads, staying clear of intensely farmed areas so that bees can collect honey and pollen free of pollutants and pesticides.

The former frontlines of the 1990s war are now among the most attractive nectar forage fronts. Having fled the warfare, only a few former residents have returned to the interstate border areas where economic opportunities after the war have become scarce. Once ravaged by conflict and still littered by landmines, the country's edgelands are overgrown with vigorous plants, invasives, and melliferous trees, which thrive undisturbed by development or agricultural schemes. During the blossoming seasons, small batches of colorful hives are parked in itinerant beekeepers' favorite nooks at a safe distance from landmine warning signs, and the land is abuzz with foraging bees. When the weather is right, honey flows along the new forage fronts.

To secure a welcome for their hives, the visiting beekeepers, Bosnian Muslims, forge working relationships with the areas' inhabitants, either the Bosnian Croats, predominantly Catholic, or Bosnian Serbs, who are mostly Orthodox Christians. To thank them for hosting the hives, the beekeepers treat the residents to the honey they harvested. When residents are elderly, ill, and lonely, as so often happens to be the case on the forage fronts, beekeepers provide seasonal care, bring medicine, run errands, and, for the length of the flow, keep company. Multiethnic, interfaith relations, ranging from tense to tender, that spring up across the forage frontlines are as brief and intense as the blossoming seasons. They do not attest to an ideal, multicultural vision of a postwar BiH any more than the flourishing frontlines foretell the country's environmental future, but they do showcase possibilities—for flow and friendship, land's reclamation, and face-to-face negotiations of differences that matter—unfostered by conservation schemes and unforeseen by formal politics.

Contemporary eco-thinkers, more generally, have taken a fresh look at landscapes wrecked by extractive and industrial enterprises or military operations to point out novel ecologies that are emerging through wreckage despite the odds.[39] Eco-stories told from within deeply disturbed grounds depart from "doomsday scenarios" to inspire readers with glimpses of alternative futures.[40]

Stories in this book, however, describe a different genre of hope, grounded in awareness of finitude and expectations of death and the world's catastrophic end.

### Listening: An Art of a Living Heart

My closest local interlocutor and teacher on the subject of Islamic cosmology and eschatology was Shaykh Ayne, a retired imam and elder of the Naqshbandi Sufi order. An imam is the one who leads the ritual prayers and, more broadly, someone who is trained and competent to lead members of a community in the formal matters of Muslim creed and conduct, the way of Islam, commonly referred to as "the straight path." Shaykh is an honorific for an elder or teacher blessed with the power of insight and endowed with the order's permission to guide others along the fast track to knowing and adoring God. The English word *order* is a rough term for Sufi "associations," which in Arabic are called tariqa (*tarkiat* in BSC), meaning a "method" and a "way" treaded by those who are drawn to a particularly contemplative and felt form of devotion.

Shaykh Ayne was an immersive storyteller. He rendered canonical sources, the Qur'an, the Hadith, and their commentaries into memorable wisdom tales. Through stories, he taught about Islam the way his Sufi elders and his father, an imam, taught, by transforming the textual tradition into an oral culture.

Sufism in scholarly literature is sometimes discussed as a discrete field of practice, an esoteric, ascetic, mystical, or ecstatic strain of devotion apart from mainstream Islam or too lettered for ordinary folk. On the contrary, Sufi stories (and songs and poetry) that circulate within tariqa are meant to reach and move the diverse listeners gathered among its members and sympathizers. Moreover, wisdom stories are well loved by Bosnian Muslims and travel widely beyond Sufi gatherings. The fact that Shaykh Ayne, like many of his teachers, spent adult life as both a Sufi dervish and an imam speaks to the close associations between sharia, the letter of the law, and the heartfelt quest for divine closeness and insight. It also suggests that storytelling is more generally a method of conveying Islamic teachings and manners through a Muslim community.

Shaykh Ayne's stories painted more than human cosmologies in which humans coexist with animals, plants, angels, and jinn, and in relationship to things and elements that were matters of God's concern. The point of telling the stories was to build the listener's character, inspire their devotional practice, and invigorate human sense of care and responsibility.

The first time we meet, Shaykh Ayne asks me whether I know what Sufism is. I tell him I have a rough idea: it is a way of practicing Islamic metaphysics. Shaykh smiles in response, then tells the following story.

A young dervish travels to a city market far from his humble dwelling in the mountains. He buys a bunch of dates. The way back home draws out under the sun's harsh rays. At long last, the dervish arrives home, covered in dust and sweat, draws fresh water from the spring, and sits down to a feast. He unwraps the dates, and there, wandering amid the glistening reddish-brown fruits is a single ant. The dervish wraps up the bundle, straps on his sandals, and heads back to return the ant to the market. "So that the ant can search for his community," Shaykh explains. "That's the living Sufism."

Shaykh's teachings were brimming full of ecological undertones. The caring way in which Shaykh Ayne related to plants, insects, and birds on our outdoor excursions and the prayerful moods he would fall into when faced with wind, sun, or rain I took as signs of his living Sufism. The degree of reverence and the breadth of contemplation were certainly particular, but the general eco-sensibility was something I could relate to. The subject of eschatology, to which our conversations returned in one way or another, was more difficult to grasp the way Shaykh intended it.

To really ponder death, to sit with it, to feel the dread of being caught unprepared, to live as if the apocalyptic event could swipe you off your feet *in a blink of an eye or sooner still* is easier said than done. There is a gap between knowing something and taking it in, to the point that it shakes you at the core and restrings your relationships with the self, with God, with all.

For that to happen, it was not enough to ponder the stories. I had to be taught to listen differently and to relate to the Qur'an, which, for Shaykh, was the ultimate reference. At least, that sort of training was implied—it took me a while to realize—in a visiting relationship with a Shaykh.

Long-term field research, which is the hallmark of an ethnographic method, often entails close working ties with people whom anthropologists consult in order to grasp something of vital importance in the field they study. For this anthropologist, visiting Shaykh Ayne was a sort of a professional friendship. On the other hand, a Shaykh is an experienced guide with teaching methods of his own and a lesson plan that amounts to nothing less than bringing his listeners and students closer to God.

Our association, thus, began with a misunderstanding. I arrived to Shaykh's home, expressing an interest in learning about "living Sufism," by which I meant the way Sufis engage with and experience that which profoundly defines them and which they hold dear. This is how anthropologists typically speak of a "living" practice. For Shaykh Ayne, however, living Sufism meant an ongoing quest for enlivening the very human subjects. I was being an academic, bent on doing research and Shaykh was living a life of one searching for divine encounter.

In the years to come, while I was duly taking notes, Shaykh was trying to break my habit of casual listening. His lessons struck me, finally, when at the tail end of 2020 this precious friend passed away.

## *Bee Caring*

While doing this research project, I kept bees on my family land. We are still at it. Trials at our apiary are not related in this book's pages, but keeping bees and caring for the land were integral to writing this book. Keeping the bees, for whom I care dearly, like keeping company with Shaykh Ayne, made me far more attentive to the signs of our times. Caring for the hives, which is becoming more difficult each season, makes me feel more vulnerable, as my home and the hives sit side by side under the same unhinged skies.

The small apiary is in an orchard on our father's land on the slope of a lowland mountain in northeastern Bosnia. Had it not been for a spectacular medieval Bosnian fort chiseled into a cliff that attracts regional tourists, the road leading to our village would be far less traveled. The population here, as in villages across Bosnia, is dwindling, as people are marrying out, migrating to other Bosnian towns or still farther north, to the European Union labor markets.

The road dead-ends at the mountaintop and forks to our hamlet at about seven hundred meters above sea level. It runs by a mosque on whose side leans a *turbe*, a small structure built around the tomb of a seventeenth-century imam and a Sufi Shaykh. Tombs of those who are said to blessed with God's friendship are found across Bosnia, and people travel from afar to deliver their prayers, leave alms, and on these auspicious sites, petition God for their wishes and troubles. Off the road, just short of where the asphalt turns into gravel, a steep, brief ride through shrubbery delivers one to the flat foothold on our land that rises up in slopes suddenly and fans out like a funnel.

Along with two small forests, the orchard is all that remains of our father's land-rich parents' inheritance. Landholding in this region was somewhat spared the early attempt at socialist collectivization in the 1950s, but it was soon lost in the fast exchange of values that took place when the predominantly rural BiH hastened onto the Yugoslav path of modern development.

Our father was among the village's early messengers of the new modern world, which was far from our hamlet but not out of reach for the mobile and the brave. He cashed in lands, forests, fields, and orchards to purchase a Vespa motorcycle and also, perhaps (our mother doubts it), to pay for his university studies. He must have looked smart on the motorbike.

Like many of his contemporaries, the professionals and new urbanites, he did not so much turn back on the land as he turned nature into a playground. Hiking, biking, and picnicking were a part of our parents' well-traveled lifestyle, which uprooted feelings for any particular place and sought in a leisurely manner nature in general. Or else, pursued "culture" across Eastern and Western Europe, that Yugoslav professional salaries and well-received passports made accessible, for a while.

Once the land was sold, my father moved to the city and his aging mother followed along. A strong woman who would not complain unless it made a difference quietly endured the city winters and, with the first spring thaw, fled to her mountain village, rented rooms to stay, and worked the remaining family land through the frostbitten fall. It was only her determined seasonal returns that forced our youthful parents to pay the land a visit back in the day. But all that changed.

After the village soil gave my father's mother all-seasons residence, the loss bent my parents to the orphaned loam, and they took to it. Over the years, they built a perfect retreat in an imperfect weekend cottage on which every door squealed and not a single one closed properly. The summer kitchen shack's door actually does close with the aid of a shoestring.

Weekend houses in the 1980s were the vogue among the Bosnian Yugoslav urbanites who grew disillusioned with industrialization and weary of modern, hectic lifestyles. Country retreats, not coincidentally, became more appealing in the wake of major industrial accidents in Tuzla City's chlor-alchaline complex in 1980 and again in 1985, followed in 1986 by the fears of the fallout radiation of "Chernobyl clouds." Meanwhile, the economy unraveled as the cycle of financial crisis shook Yugoslavia, which was never to recover, and introduced supply chain shortages, currency devaluations, and subsistence anxieties that unnerved many socialist moderns.

In pursuit of healthy foods and wholesome lifestyles, and along with scores of other Bosnians of their generation who had the means, my parents tracked "back to nature" to my father's native village for holidays and weekends, grew food, picked mushrooms, preserved berries, and overplanted trees and flowers. Today's orchard is a thriving inheritance of past relationships to land, including our father's father, a grafter without a peer who could make two different apples grow out of a single tree. The trees bear his presence better than the single faded photo he left behind before his trace was lost on the northern fronts of the Second World War.

During the 1990s war in Bosnia, our village was on the safer side of the combat zone. During the hungriest, coldest years for the town of Tuzla, sieged and shelled, the village was a shelter for the four of us: mother; Azra, the youngest of the three sisters; Brenka the goat; and me.

When the war was over, our parents retired from defunct state companies, and, after a few failed attempts at private enterprises, they moved to the village. First, like my grandmother, they resided in the mountain home from thaw to frost, then for good. They went to the city only reluctantly, to see physicians, do banking, or visit family and friends. While our mother made rounds, our father stayed by the car parked on the roadside and just about kept the engine running, ready to flee. And so, for years.

I traveled back from the States to spend summers doing research in Bosnia and spend time in the village. Azra traveled from Cuba on a break from filmmaking school. Summers gathered us in the village on vacations. When the research project on beekeeping started, we built an apiary. That apiary, we quickly learned, committed us to the land year-around, with tasks of care that someone had to pick up while I was abroad. The bees' needs rounded and grounded our family through seasons.

My mother stayed on after my father passed away in 2016. Generous research grants afforded me longer spells in the field and by the apiary. With Azra back from film school we started making a documentary film about beekeeping and Muslim end time lore. I began dreaming of a life by the bees on our mountaintop and, eventually, planned to build a tiny house on our land.

"This is a disaster in the making," the building engineers we invited, pronounced, surveying the steep sides of our land. Everywhere, they saw signs of landslide movement: surface waters and underground streams, cracked and furrowed soil, and leaning trees. There were wicked forces, deep below the grass, brewing our land's imminent collapse, the engineers said. Sure, all right, a disaster in waiting. But not only that. Fluid is the very ground of all our whereabouts, no matter how paved and grounded, in the *ākhir al-zamān*, the times of global environmental disasters and climate change. At the spot that a geodetic engineer evaluated as the soundest, we decided to build.

From the top of our land, you can see far along the valley below as it folds into mountains after mountains along the horizon. You can also see rapid changes taking place closer below: the town and villages are swelling up and burning coal, which is still the primary means of heating and burning car fuel. You can see mountains ripped apart and machine-eaten in the distant quarries. And far away but not far enough from anyone, sickly emissions are rising from Tuzla's coal-based power plants. From this vantage point, at the place that fed us and kept us warm throughout the war, the place from which the 1990s frontline shelling was barely audible as a distant noise, we learned in the course of these beekeeping years just how exposed we are to bad weather.

Unpromising climate is a sign of our times, beyond Bosnia. There is no place high enough or far enough to afford a refuge, although there are different vantage points; from some, the climate change appears manageable, from others, our planet's future appears unlivable.

Learning to listen to Islamic eschatology has been a search for the vantage—or the mood—from which to think, live, and care while doing this research. If you threw a rope to the deepest ground, a hadith says, it would fall upon God. God is all encompassing, the prophetic saying suggests, al-Wāsi' (the Vast) being among His divine names. Divine attributes that the Qur'an particularly describes as all embracing are mercy and knowledge. Starting from this premise, there is no room in Islamic perspective for despair. Nor can a careful listener of the Qur'an afford to be unconcerned about the state of the troubled world and one's hand in it, or think himself or herself excepted from the warnings. The whole point about watching for the signs of the bad end is to work—and pray—wholeheartedly against it. The Qur'an quotes God: *Soon I will show you My signs, so do not hasten Me.*

### Notes

1. 59:19. Throughout the book, I provide translations of excerpts from the Qur'an that aim to convey something of its literary quality while also trying to stay as close as possible to the meanings grasped by authoritative translations and commentaries. My knowledge of Arabic is inadequate for the task, and so I have leaned on commentaries given to me by Shaykh Ayne and other local experts while making use of classic translations by A. J. Arberry, Abdullah Yusaf Ali, Mohsin Khan, and Muhammad Pickthall, as well as the recent translation under the chief editor, Hossein Seyyed Nasr. In addition, I have consulted the translation to BSC in Besim Korkut's *Kur'an s prijevodom* (Medina, Saudi Arabia: Hadimu-l-Haramejni-š-Šerifejni-l-Melik Fahd, 1992).

2. For readings of pop-apocalyptic genres in the contemporary Muslim world, see David Cook, *Contemporary Muslim Apocalyptic Literature* (Syracuse: Syracuse University Press, 2008) and Jean-Pierre Filiu, *Apocalypse in Islam*, trans. M. B. DeBevoise (Oakland: University of California Press, 2011). Jamel A. Velji develops an interesting argument for the salience

of the "apocalyptic" in Islamic history in Velji, *An Apocalyptic History of the Early Fatimid Empire (Edinburgh Studies in Islamic Apocalypticism and Eschatology)* (Edinburg: Edinburgh University Press, 2016). The most insightful and comprehensive overviews of Islamic eschatology are the following: William C. Chittick, "Muslim Eschatology," in *The Oxford Handbook of Eschatology*, ed. Jerry L. Walls (Oxford: Oxford University Press, 2008), 132–150; William C. Chittick, "Eschatology," in *Islamic Spirituality: Foundations*, ed. Seyyed Hossein Nasr (New York: Crossroad, 1997), 378–409; Hamza Yusuf, "Death, Dying, and the Afterlife in the Qur'an," in *The Study Qur'an: A New Translation and Commentary*, ed. Hossein Seyyed Nasr, Caner K. Dagli, Maria Massi Dakake, Joseph E. B. Lumbard, and Mohammed Rustom (New York: HarperCollins, 2015), 1819–1855; Jane Idleman Smith and Yvonne Yazbeck Haddad, *The Islamic Understanding of Death and Resurrection* (Oxford: Oxford University Press, 2002); and Sebastian Günther and Todd Lawson, eds., *Roads to Paradise: Eschatology and the Concepts of the Hereafter in Islam* (Leiden: Brill, 2017).

3. 50:19–20.
4. 33:45.
5. See chapter 4 of this book.
6. With a nod to Donna Haraway, who has written the most compelling call for making kin across species. Donna Haraway, *When Species Meet* (Minneapolis: University of Minnesota Press, 2008).
7. As Jürgen Tautz puts it, the colony changes with each queen like a "genetic chameleon." Jürgen Tautz, *The Buzz about Bees: Biology of a Superorganism* (Berlin: Springer, 2008), 45.
8. Thomas Dyer Seeley, *Honeybee Democracy* (Princeton, NJ: Princeton University Press, 2010) and *The Lives of Bees: The Untold Story of the Honey Bee in the Wild* (Princeton, NJ: Princeton University Press, 2019).
9. 6:38. The original Qur'anic Arabic, *dābba*, is more expansive than the English "animal" or Bosnian-Serbian-Croatian 'životinja' would suggest. See Sarra Tlili, *Animals in the Qur'an* (New York: Cambridge University Press, 2012), 71.
10. 16:68–69.
11. *Our Lord is He who gave each thing its form then guides it*, 20:50.
12. Therapeutic uses of hive products, known under the international label of "apitherapy" is sometimes dismissed in a knee-jerk fashion, as in Mark Winston's otherwise resourceful book on the wide range of meanings of the honeybee. See Mark Winston, *Bee Time: Lessons from the Hive* (Cambridge, MA: Harvard University Press, 2014), 142–148. Winston calls api-therapy "cultish" and attributions of medicinal properties to bee products merely anecdotal. On the contrary, api-therapy is guided by the principles of modern evidence-based science and comprises a massive field of international and interdisciplinary experts who are publishing, in English, results of research conducted mostly outside of Euro-American knowledge production centers. Among the wide range of journals publishing studies on medicinal uses of hive products are *Trends in Food Science and Technology*, *Arabian Journal of Chemistry*, *International Quarterly Journal of Research in Ayurveda*, and *Asian Pacific Journal of Tropical Biomedicine*. In 1971, Apimondia, the International Federation of Beekeepers' Associations, which gathers annually for a meeting of professionals, bee scientists, and commercial, UN, and nongovernmental partners, organized a symposium on the uses of bee products in human and veterinary medicine. Apitherapy from then on was granted a formal venue within Apimondia, while the establishment of a Standing Scientific Commission on Apitherapy in 1983 provided an axis for annual gatherings, transnational research collaborations, and an exchange of research data and concerns. Bosnian journalist and beekeeper Nijaz Abažić closely followed Apimondia proceedings and is largely responsible for popularizing api-therapy in Yugoslavia, starting with his 1967 *The Secrets of Bee's Honey*. Nijaz Abadžić, *Tajne Pčelinjeg Meda* (Sarajevo: NIP "Zadrugar," 1967).

13. 41:44; 10:57.
14. Sarajevo-based artist, Meliha Teparić has inscribed this water fountain for the bees with Qur'anic verses on the honeybee, as a part of an art installation project that we planned at our village apiary.
15. Ismaīl Ibn Kathīr, *The Signs Before the Day of Judgment* (London: Dar Al Taqwa, 1997), 7.
16. Ibid., 6.
17. 16:77.
18. 7:187.
19. 79:42–43.
20. Imam Birgivi's text was on the school curriculum across the Ottoman Empire and still used in early twentieth-century Bosnian madrasa, during Shaykh Čolić's schooling. Mustafa Čolić, *Et Tarikatul Muhammedijjetul Islamijjetu: Evidencije i definicije islamskih šerijatskih učenja i vjerovanja* (Visoko: Kaligraf, 1998). Čolić's translations and commentaries of other parts of Birgivi's work were prepared for publication by his dervishes, after Shaykh's passing, in two books: Mustafa Čolić, *Et-Tarikatul-Muhammedijjetul-Islamijjetu, učenje i moral Allahovog Poslanika Muhammeda a.s.: Srčano zdravlje i bolesti metafizičkog insana* (Visoko: Tekija Šejh Husejn-baba Zukić, 2016) and Mustafa Čolić, *Et Tarikatul Muhammedijjetul Islamijjetu, učenje i moral Allahovog Poslanika Muhammeda a.s.: Zdravlje i bolesti jezika i ostalih organa metafizičkog insana* (Visoko: Tekija Šejh Husejn-baba Zukić, 2020).
21. Čolić, *Et Tarikatul Muhammedijjetul Islamijjetu: Evidencije i definicije*, 264, 267.
22. Ibid., 268. Čolić has referred to objective things and events as *išaret* (from Arabic *ishārah*), which is a technical term in Sufi thought for meanings gained at the level of a deeply mindful heart by means of contemplation and inspiration. I translate freely from Čolić's books, hoping to convey the gist of his arguments, although my discussions do not delve deeply enough into Čolić's work, which is exceptionally complex and articulated in the language that is quite his own.
23. Shaykh Čolić has translated, with comments, al-Ghāzalī's famous discussion on the heart. Mustafa Čolić, *Zagonetnosti i Nepoznanice Metafizičkog Srca (Metafizičkog Čovjeka)* (Visoko: Tekija Šejh Husejn-baba Zukić Hukeljići-Živčići, 2000). For most insightful considerations of the heart in Islamic spiritual tradition, see Sachiko Murata, *The Tao of Islam: A Sourcebook on Gender Relationships in Islamic Thought* (New York: State University of New York Press, 1992), 289–320; William Chittick, *Divine Love: Islamic Literature and the Path to God* (New Haven, CT: Yale University Press, 2013).
24. Shaykh Ayne used to relate a saying attributed to the Prophet: "One's *nafs* [soul] is more dangerous than seventy devils."
25. Although ecology does not readily come up in mainstream conversations on Islam, the Quran and the Hadith lavish exceptional attention on plants, animals, and the elements, as a number of scholars of Islam and Muslim intellectuals and environmentalists have argued. The core texts of Islam recommend deep concern for animal welfare and promote planting, fruit farming, nature conservation, and careful use of resources. At the same time, environmental ruination is intensifying in predominantly Muslim countries as anywhere else in the modern world, and Islamic tradition is rarely the framework that guides animal farming or environmental management in Muslim majority statea. Seyyed Hossein Nasr, an outstanding contemporary thinker and interpreter of Islamic and Sufi sources, has long been arguing that the environmental crisis is also the spiritual crisis of modern humans, Muslims included. The argument is familiar and still profound even in this restatement: once nature is profaned, once it is reduced from the domain of Divine Self-Manifestation, whereby every leaf reflects the Face of God, to extractable resources and inert, nonhuman entities, humans are locked out of a meaningful universe. The cosmos shrinks to the measure of the human intellect and the scope of lone human imagination, which for all their glory are limited. In addition,

overemphasis on narrow politics by Muslims and commentators on Muslim affairs alike has overshadowed most other dimensions of Islam. See Seyyed Hossein Nasr, *Knowledge and the Sacred* (Albany: State University of New York Press, 1989), and *Man and Nature: The Spiritual Crisis in Modern Man* (Chicago: Kazi, 2003); C. Richard Foltz, Frederick M. Denny, and Azizan Baharuddin, eds., *Islam and Ecology: A Bestowed Trust* (Cambridge: Center for the Study of World Religions, Harvard Divinity School, 2003); Richard C. Foltz, *Animals in Islamic Tradition and Muslim Cultures* (London: Oneworld, 2005); Mawil Izzi Dien, *The Environmental Dimensions of Islam* (Cambridge: Lutterworth, 2000); Fakhar-i- Abbas, *Animal's Rights in Islam: Islam and Animal's Rights* (Riga: VDM Verlag, 2009).

26.  For excellent examples, see Human Animal Research Network, *Animals in the Anthropocene: Critical Perspectives on Non-Human Futures* (Sydney: Sydney University Press, 2015); Stacey Alaimo, *Bodily Natures: Science, Environment, and the Material Stuff* (Bloomington: Indiana University Press, 2010), *Exposed: Environmental Politics and Pleasures in Posthuman Times* (Minneapolis: University of Minnesota Press, 2016), and "Elemental Love in the Anthropocene," in *Elementary Ecocriticism: Thinking with Earth, Air, Water, and Fire*, ed. Jeffrey Jerome Cohen and Lowell Duckert (Minneapolis: University of Minnesota Press, 2015), 298–309; T. J. Demos, *Decolonizing Nature: Contemporary Art and the Politics of Ecology* (Berlin: Sternberg, 2016); Donna Haraway, "Otherworldly Conversations, Terran Topics, Local Terms," in *Material Feminisms*, ed. Stacey Alaimo and Susan Hekman (Bloomington: Indiana University Press, 2008), 157–187; Anna Tsing, "Blasted Landscapes (and the Gentle Arts of Mushroom Picking)," in *The Multispecies Salon*, ed. Eben Kirksey (Durham, NC: Duke University Press, 2014), 87–110; Anna Tsing, "Unruly Edges: Mushrooms as Companion Species," *Environmental Humanities* 1, no. 1 (November 2012): 141–154, https://doi.org/10.1215/22011919-3610012; Jeffrey Jerome Cohen and Lowell Duckert, "Introduction: Eleven Principles of the Elements," in *Elemental Ecocriticism: Thinking with Earth, Air, Water, and Fire*, ed. Jeffrey Jerome Cohen and Lowell Duckert (Minneapolis: University of Minnesota Press, 2015), 1–26; Mel Chen, *Animacies: Biopolitics, Racial Mattering, and Queer Affect* (Durham, NC: Duke University Press, 2012).

27.  Marisol de la Cadena, "Indigenous Cosmopolitics in the Andes: Conceptual Reflections beyond 'Politics,'" *Cultural Anthropology* 25, no. 2 (April 2010): 334–370, https://doi.org/10.1111/j.1548-1360.2010.01061.x; Anders Blok, "War of the Whales: Post Sovereign Science and Agonistic Cosmopolitics in Japanese-Global Whaling Assemblage," *Science, Technology & Human Values* 36, no. 1 (November 10): 55–81, https://doi.org/10.1177/0162243910366133; Stacey Ann Langwick, *Bodies, Politics, and African Healing: The Matter of Maladies in Tanzania* (Bloomington: Indiana University Press, 2011).

28.  See Bruno Latour, *Facing Gaia: Eight Lectures on the New Climatic Regime*, trans. Catherine Porter (Cambridge, UK: Polity, 2017), 218.

29.  A "new age of curiosity" is a term I lift from Michelle Foucault's wistful 1980s vision of the coming age that would reinstate bolder concern for "what exists and what could exist" and "regard otherwise the same things," and I repurpose it as a shorthand for a host of contemporary theoretical ventures that, jointly, have broadened the investigative range of the social and humanities. Foucault's complaint about the stifling of Western social thought is still current (see Peter Skafish, "Anthropological Metaphysics/Philosophical Resistance," Theorizing Contemporary, Fieldsights, posted on January 13, 2014, https://culanth.org/fieldsights/anthropological-metaphysics-philosophical-resistance), but the inquiries into "experience" are currently "thinking paths and possibilities" (see Michael Foucault, *Politics, Philosophy, Culture: Interviews and Other Writings 1977–1984*, trans. Alan Sheridan [New York: Routledge and Chapman & Hall, 1988], 328). The new age of curiosity goads inquiry beyond human to animal, vegetal, and microbial lives. Animal and animality studies poke at the former certainties about the human-animal distinction while multispecies ethnographies flesh out ecologies of species companionships and coconstitutions under anthropogenic pressures. Turning to

the matters that previously sat idle and low in the old hierarchies of animacy has been even more daring, opening up things at hand as well as liberally defining nonhumans, as Isabelle Stengers says: "whatever forces thought" (see Isabelle Stengers, "Reclaiming Animism," *e-flux Journal* 36 [July 2012]: 7, https://www.e-flux.com/journal/36/61245/reclaiming-animism/). The newly spurred interests in nonhumans frequently bring together politics and metaphysics, asking again, "What can be known?" and "What matters?" New curiosity is recommended as self-transformative. Stengers suggests we shed the "sad, monotonous little critical or reflexive voice" and stay open "to wonder" (Mary Zournazi and Isabelle Stengers, *Hope: New Philosophies for Change* [New York: Routledge, 2003], 244–272). Latour, in particular, is politically minded when he says it is high time to "go back to the old question of 'what is X?'" (Bruno Latour, *An Inquiry into Modes of Existence: An Anthropology of the Moderns*, trans. Catherine Porter [Harvard: Harvard University Press, 2013], 21). He opens inquiry into modes of existence, to populate the cosmos with a diversity of beings, granting them a substantial reality within a "richer ecosystem" (Latour, *An Inquiry into Modes of Existence*, 11, 18). The way to negotiate the Pluriverse, however, is to forge a minimalist metaphysics (Bruno Latour, *Politics of Nature: How to Bring the Sciences into Democracy*, trans. Catherine Porter [Harvard: Harvard University Press, 2004], 61). Multispecies and animal studies are more interested in earthbound ontologies, with Donna Haraway's pathbreaking work, her often-cited praise for the "mortal world-making entanglements" (Haraway, *When Species Meet*, 4) and the gamey proposition that *posthumanities* is "another word for 'after monotheism'" (Haraway, *When Species Meet*, 245). An exception is Anat Pick's book, which aspires to a "rapprochement between the material and the sacred," even if Pick's text does not quite show the way (Anat Pick, *Creaturely Poetics: Animality and Vulnerability in Literature and Film* [New York: Columbia University Press, 2011], 17). With a similar ambition, a volume edited by Stephen Moore is exploring the concept of "creaturely theology" (Stephen Moore, ed., *Divinanimality: Animal Theory, Creaturely Theology* [New York: Fordham University Press, 2014], 11).

30. In the words of environmental philosopher John Foster, the "secular, naturalistic general picture is now (despite various forms of rearguard religious protests against it) the working worldwide image of the advanced societies, and it is implicit in the terms in which these societies have globalized themselves through science, technology, telecommunications, and the capitalist-individualistic economic model. It is the human self-recognition, we might say, that keeps the internet functioning." This thinker who is otherwise deeply critical of the modern culture that is driving the planet to the edge assumes that religion can only speak from the fringes in the tone of protest and that globalization has achieved a consensual human self-perception. John Foster, *After Sustainability: Denial, Hope, Retrieval* (New York: Routledge, 2015), 118–119. See also Jean-Luc Nancy, *After Fukushima: The Equivalence of Catastrophes*, trans. Charlotte Mandell (New York: Fordham University, 2015), 35.

31. A noteworthy example is a volume edited by Stephen Moore, which starts from the premise that the expanded range of inquiry beyond human has already set conditions for rethinking the role of the divine. Moore, *Divinanimality*. See also Anat Pick's suggestive *Creaturely Poetic*.

32. Zournazi and Stengers, *Hope*, 244–272.

33. Shaykh Čolić's books are an extended commentary on the sort of knowledge that is in decline: Mustafa Čolić, *Et Tarikatul Muhammedijjetul Islamijjetu: Zbirni Ilmihal islamizacionih stanja i pitanja za odrasle i dorasle, Ilmihal za Odrasle i Dorasle* (Visoko: Tekija Šejh Husejn-baba Zukić, 2000), 224, 231.

34. 55:29.

35. Throughout his works, Čolić uses Qur'anic Arabic terminology to develop an analytical language that is conversant with philosophy but thoroughly indebted to the Revelation and is, at times, lucidly vernacular. His discussion of forms of divine self-revelation is synthesized in *Kelamske i Tekvinske Božanske Obznambene Objave i Pojave i Njihovi Kira'eti (Čitanje i Učenje)* (Visoko: Tekija Šejh Husejn-baba Zukić Hukeljići-Živčići, 2003).

36. For an overview of phenological studies to date, diversity of species responses, and methodological challenges entailed in grasping the current and future trends, see Camille Parmesan, "Influence of Species, Latitudes, and Methodologies on Estimates of Phenological Response to Global Warming," *Global Change Biology* 13 (2007): 1860–1872; Stein Joar Hegland, Anders Nielsen, Amparo Lazaro, Anne-Line Bjerknes, and Ørjan Totland, "How Does Climate Warming Affect Plant-Pollinator Interactions?," *Ecology Letters* 12 (2009): 184–195; Camille Parmesan, "Range and Abundance Changes," in *Biodiversity and Climate Change: Transforming the Biosphere*, ed. Thomas Lovejoy and Lee Hannah (New Haven, CT: Yale University Press, 2019), 25–38; Eric Post and Michael Avery "Phenological Dynamics on Pollinator-Plant Associations Related to Climate Change," in *Biodiversity and Climate Change: Transforming the Biosphere*, ed. Thomas Lovejoy and Lee Hannah (New Haven, CT: Yale University Press, 2019), 42–54.

37. See Thomas Lovejoy and Lee Hannah, eds., *Biodiversity and Climate Change: Transforming the Biosphere* (New Haven, CT: Yale University Press, 2019); Jedediah Brodie, Eric Post, and Daniel F. Doak, *Wildlife Conservation in a Changing Climate* (Chicago: University of Chicago Press, 2013).

38. Textual references that relate good faith with hope and cheer are many. Most commonly cited in local conversations are Qur'anic verses "my Mercy encompasses all things" (7:156) and "do not despair of God's comfort. Indeed no one despairs in God's comfort except the unbelievers" (12:87).

39. Eben Kirksey, *Emergent Ecologies* (Durham, NC: Duke University Press, 2015); Tsing, "Blasted Landscapes," 157–187; Bridget Guarasci and Eleana J. Kim, "Ecology of War," Theorizing Contemporary, Fieldsights, posted January 25, 2022, https://culanth.org/fieldsights/series/ecologies-of-war.

40. "Contemporary writing on the environment is largely focused on doomsday scenarios," Eben Kirksey notes in the introduction to the book that grounds reasons for hope in disturbed landscapes and unscripted multispecies affinities. Kirksey, *Emergent Ecologies*, 6. Scholars writing about animal extinction and endangerment are experimenting with new and older narrative styles and affects, including nostalgic, utopian, contemplative, and even comic, to move the audience to register, mourn, or counter damages and losses that are becoming routinized: Hayden Fowler, "Epilogue New World Order—Nature in the Anthropocene," in *Animals in the Anthropocene: Critical Perspectives on Non-Human Futures*, ed. Human Animal Research Network Editorial Collective (Sydney: Sydney University Press, 2015), 243–254; Thom van Dooren, *Flight Ways: Life and Loss at the Edge of Extinction* (New York: Columbia University Press, 2014); Deborah Bird Rose, *Wild Dog Dreaming: Love and Extinction* (Charlottesville: University of Virginia Press, 2011); Stacey Alaimo, *Exposed: Environmental Politics and Pleasures in Posthuman Times* (Minneapolis: University of Minnesota Press, 2016). Philosopher of science Isabelle Stengers thinks that "we have a desperate need for other stories" and calls for tales about achievements, however small, that would reseed "the devastated desert of our imagination." Isabelle Stengers, *In the Catastrophic Times: Resisting the Coming Barbarism*, trans. Andrew Goffey (London: Open Humanities with Meson, 2015), 132.

# − 2 −

# *Honey's on the Wheels*
## Beekeepers' Prayers

Dawn breaks. Light filters through the forest, softened by the morning fog. Jusuf and Nijaz look over the thirty hives they off-loaded in the dead of night. Pleased with a job well done, the beekeepers turn to head out of the forest, then freeze in their steps. Well ahead of them, newly visible, is the back of a bright red sign, standing tall with dark advice. The previous night, they had stepped unknowingly beyond the land mine warnings that fringe these former frontlines. Nijaz laughs nervously. Jusuf silently prays. They walk in single file, sticking closely to the trail they cut into the shrubbery last night when, under the narrow streaks of flashlights, they walked back and forth from their beaten-up van cradling hives in their arms. Lucky fellows, they get by unharmed.

Manda awaits them with strong Turkish coffee just off the stove. For the length of the black locust flow, this elderly woman who lives nearby will be their bees' host. Feeling raw from a sleepless night and still jittery from walking through the land mines, the men are silent, their palms wrapped around small, piping-hot coffee cups. Manda is chattering, thrilled to have them back, and like a good Bosnian host, she frets over their comfort: "More coffee? Have some cookies. Have you warmed up?"

For much of the year, Manda lives a lonely life in a Bosnian Catholic village ruined by the war. Residents fled the fighting in the 1990s, and only a few, mostly elderly people, have returned to this borderline between the two entities of the ethnically divided state of Bosnia and Herzegovina (BiH). Battle-scarred, the area known as Posavina at the basin of the Sava River is rewilding. Overtaken by native and invasive species, this is one of many, war-ravaged, neglected places across the country that turned into honeyed heavens.

Black locust trees, fast and eager growers, have spread and thickened in the mild, riverine climate. Manda keeps a watch on the trees each spring, as eagerly as beekeepers do across the country. From within the thorny branches, tiny buds explode in long, hanging clusters of curled blossoms, breezily fragrant, and transient to a fault. For roughly three weeks, heavy, bluntly sweet nectar will flow before florets darken, sag, and drop. The black locust foraging season is known to be brief and stormy, and for the length of it, Manda will enjoy the company of visiting bees and their Bosnian Muslim beekeepers. While bees are being matched to their partner plants, forage frontlines foster friendships between visiting beekeepers (Bosnian Muslims) and their hosts (Bosnian Catholic Croats and Bosnian Orthodox Serbs) in areas that are too marginal for formal interethnic reconciliation.

Scouter bees are already out investigating. Foragers will shortly follow their danced cues to the pollen and nectar. The black locust season begins, and, God willing, honey will flow.

### *Forage Fronts*

Picture this: It is springtime, and the thickets of eastern and northeastern Bosnia are bursting out with ringlets of snow-white black locust. The thorny shrub that qualifies as an invasive has subtly reclaimed the former frontlines of the 1990s war, seeded by land mines and left fallow by residents that fled the combat violence as much as the economic underdevelopment since the peace. Mobile beekeepers from across BiH bring hives in small batches—anywhere from thirty to a few hundred—to their favorite coves.

"Honey's on the wheels," local beekeepers say. Summed up in the saying is a complaint about the hard times and the gist of their coping strategy. As honey is becoming scarce, mobile beekeepers travel, chasing after it. They tour the country's edge lands as far as possible from the coal-powered industries that massively pollute the urban and suburban environments of central and northeastern Bosnia. Likewise dodging the farmlands treated by agricultural chemicals, they search out nectar-yielding seasonal forage.

The bees that come to the black locust forests are typically strong in numbers, relatively healthy, and hungry for what is effectively the first rich nectar flow in the region. They storm the blooming trees. The beekeepers hold their breath for the length of the black locust flow. Its floral-tasting honey is not just the favorite in the region; it is also praised for its medicinal properties. Its appearance is decisive for the turn of the foraging year as a whole.

Depending on the weather, a hive can gather up to ten kilograms (twenty-two pounds) of nectar a day. Compared with earlier spring bloomers, the sugar content of black locust nectar is highly concentrated, so after the excess water evaporates, the hives are still left with plenty of honey. The black locust nectar flow feeds the population of bees, expanding at the turn of the summer. In the course of three to four weeks, blooming tree crowns are also expected to supply the visiting hives with honey food stocks and surpluses for apiarists to harvest. Not least, the flow is counted on to fuel the building of new wax comb—the vital infrastructure for the bees' reproduction and communication.

Beekeepers' expectations for black locust honey, however, now belong to the memory of a lost climate. The black locust nectar has mostly failed to yield, in every year I spent with beekeepers in the field, from 2014 to 2017, and in each subsequent years I spent visiting and filming the forage grounds through 2021.

The circumstances of the flow and its failure are infinitely particular across the sites and vary from one year to the next, given the variability of weather and the peculiarity of microclimate niches along the forage fronts.

In some places, black locust trees barely bloom or not at all. Elsewhere, the tree crowns are sagging with blossoms, but the nectar flow is spoiled by a sudden spell of bad weather. Hail, frost, and cold showers ruin the blossoms while the bees, pent up in the hives, eat quickly through their food stocks. At times, the bloom is rich and the weather seems right, but the insects are nearly starving. Unless or until something else flows, the bees must be artificially fed with sugar syrup or patties made of pollen or pollen substitutes lest they starve, swarm, or curb the brood. Supplemental feeding is an old tactic, as honeybees' apiary lives have always alternated between flow and dearth, but the frequent and prolonged need for artificial feeding, apiarists say, is entirely new.

At the first signs of diminishing nectar, mobile beekeepers move on. Traveling skillfully cross-country on fluid schedules, they bring their hives to the next promising spot, where nectar is on the cusp of emerging. Its forthcoming possibility dawns in auspicious signs—such as the tassels the black locust tosses out, signaling an intent to bloom—although the weather, that mercurial medium of honey's production, can always shift. Predicting blossoming based on environmental cues is the beekeepers' essential skill, although the phenologies—the seasonal life cycles of plants—emerging through the new habits of the weather defy conventions and best guesses.

The overall trends captured in the beekeepers' and my field notes suggest that honeybees and beekeeping lifestyles are becoming hard to sustain, despite savvy nectar predictions and nimble itineraries along the forage fronts. And yet beekeepers' plans and expectations for the honey seasons are not dampened in advance by the glum records that seem to suggest that climate change is dooming honey flow.

Tracking honey along the forage fronts, I found novel vagaries of nectar flow. Fast-paced changes to the honey ecologies overwhelmed local apiculturists but, being savvy professionals and devoted bee lovers, they managed. Keeping bees on the wheels throughout the summer, they invigorate clever, honey chasing strategies. In response to the unseasonal and fickle weather, they adjust their itineraries and scout familiar landscapes more intensely in search of the nooks that may nourish rare chances of strong flow.

But it is not just the skilled forecasting or the well-developed routes that keep the beekeepers going throughout the dearth. A prayerful anticipation that honey will flow, God willing, inshallah, accompanies their efforts and counters disappointments.

Fig. 2.1 On the forage fronts

I am interested in what this simple, casual invocation says about the Muslim beekeepers' sense of sweet chances in the honey-waning world. What does it imply about God's knowledge and divine giving?

Divine will, knowledge, and generosity are matters given their due theological and metaphysical consideration in Islamic textual sources. Nonetheless, profound meanings of such vital terms seep into day-to-day practices as they are presumed and pondered in wisdom stories, invocations, and Qur'anic references that people share. Prayers are about metaphysics getting practical not least under the conditions of dire need and deep uncertainty. More than ever, inshallah is recognized as the condition indispensable to honey hunting in the estranging atmosphere.

One could simply gloss the invocation as a case of religious utterance that says nothing more than how some believers conceive of their world. Taken in that way, invocation appears as an entirely internal affair, a mindful or emotional process that makes a difference to an individual or community experience, shaping believers' perceptions, outlooks, and actions. Prayer, by this account, is quite different than forces that a climate biologist observes in the real world of rising carbon emissions, altering biosphere, changing regimes of precipitation, or conservation efforts.

In Islamic sources, on the contrary, invocation is recommended precisely because it works on the inside and on the outside. Because prayer is a way of participating in the world, I am curious what invocations do to the people and to landscapes swept up by the climate change. What kind of relationships between God, weather, bees, plants, and human subjects do prayers invoke, restate, and put to work?

## *Moving On*

In two weeks, the black locust nectar has become a mere trickle. White, rubbery blooms still hold, fragrant and bright, but few honeybees visit the trees, which is a clear sign that the foraging prospects have closed. Jusuf and Nijaz give their strongest hives a tug to gauge their gain. The flow, judging by the weight, is modest. They must move on quickly before the hives' honey stocks depreciate.

Days earlier, they surveyed the wider area and considered their options. Along the riverbanks, farther into the flatlands, indigo bush, a later bloomer, is opening up. A tough, leggy shrub that prefers wet soils but tolerates all types of loam, it sprouts purple spike-like floral clusters flashing with gold-yellow antlers. Full of pollen, flowers are irresistible to the bees. From there, they could move back inland to linden forests come June.

Later still, in July, they could travel south to seek higher altitudes. Highland meadows and forests, with their many diverse bloomers, promise a reserved but steady nectar throughout the summer and the possibility of a honeydew surge in the late fall. These nectar flow opportunities draw beekeepers and have made the mountains famous for their honeys.

Jusuf and Nijaz's honey-hunting routes have been elaborated over the years. The overall forage map has not changed significantly, although their travels yield less each subsequent year, since I have followed these two beekeepers, from 2015 through 2021. Even with multiple possibilities sketched out in advance, the burning question of where exactly to go next can only be settled by active scouting. Plants bloom when a number of elements line up auspiciously, and they secrete nectar in the chink of a good weather spell, which is narrower than ever since the local atmosphere has become unhinged from the old customs of four seasons.

Nijaz is the one to forecast nectar flows. Over the years, he has cultivated a passion for plants: he devoutly reads horticultural manuals, studies regional plant phenologies, and has become quite a grafter. An intense man, Nijaz is a talker. A pilot who once trained with the Yugoslav Air Force before myopia disqualified him from service, he built a career as a bank manager until the insolvent bank went under in the rickety postwar financial market. Since then, when he is not with the bees on the wheels, Nijaz does odd jobs: he sells carpets or grafts fruit trees and maintains orchards for amateur horticulturists. His wife and adult children are successful professionals, while he ekes out an income in the low-wage sector. A proud man with a thick air of disappointment, he intently watches twigs and tree crowns, sifts through the ambient cues, and makes his bet on a nectar spot.

Jusuf trusts him. An easygoing man, Jusuf is more prone to appreciating the uncertainty that honey hunting entails each season. A funny man, Jusuf bears himself with an understated self-confidence. He was a special operations soldier in the 1990s war with the army of BiH. Since then, he has spent several years as an electrician in high-risk posts with military defense contractors in Afghanistan. The stories he tells. A mountaineer at heart, his posttraumatic stress disorder symptoms flare up when winter months hold him hostage in the smoggy town of Tuzla. He can lose his temper, though his wife and daughter will have none of it, so he works it off, he says, by walking the dog, planning the spring gardens, and keeping the koi fish happy.

When the nectar road opens come spring, Jusuf's nerves calm. In the evenings, he and Nijaz wait by the hives for the foragers to return. Then they close the entrances, remove the roofs so air circulates into the nests through the wire mesh ceilings, and secure the honey supers with straps. The two will load up the bees at night and drive them to the next sweet spot—once they decide which one it will be.

Their repurposed 1980s Volkswagen van holds no more than thirty hives, so it takes three trips over three days to move all the bees. Their host Manda intently follows the hustle of packing and moving from around the corner. At the thought of them leaving, she tears up. Nijaz has whitewashed the tree seedlings he planted the previous year in her yard. They brought her prescription medicine and daily groceries and took her to town to run errands. When they sit down over meals, she tells of the long winter nights she spends alone in her room by the woodstove stoking the fire. And how she prays. A devout Catholic, her small bed rests beneath a framed image of Jesus with his heart aflame. When she dies, she worries aloud, no one will be there to bury her or care for her grave.

Nijaz promises to call over the course of the year. Besides, Manda will likely outlive the two of them, he teases. She laughs and touches up her pretty ink-blue headscarf blooming with white flowers. He is her favorite of the two.

By nightfall, Jusuf and Nijaz settle on the indigo bush and agree on the particular site. If the bush flows well, the total gain per hive might be as high as thirty kilograms (sixty-six pounds). On top of that, bees build brood and wax well on its plentiful pollen.

In May, even if the black locust has failed, the honey season will still be young and the chances fresh. Within three days, the last hives are loaded. Manda stays up late to say goodbye. The beekeepers drive away. Leaning on a cane, she walks into the house to rest.

At the end of the summer, Jusuf wraps up the year for me over the phone: a disaster. Like the black locust, the indigo bush also failed; the flowers came and went while the bees stayed indoors through cold, long-lasting rains. The linden did not fare much better in the extreme heat that followed. The two beekeepers harvested a "symbolic quantity" of honey just to have a taste.

"We're still hoping for honeydew honey, though," Jusuf says, "God willing." Honeydew, or manna, as it is also known, is the sweet substance secreted by aphids or scale insects gorging on plant sap, which bees collect to make honeydew honey. It is extremely rare. God willing, nonetheless.

### *Grounds for Hope*

The saying "God willing" invokes the very possibility of manna flowing. Manna, a type of flow that is marvelously bountiful as well as incomparably unpredictable and rare, is an appropriate flavor of an invocation that anticipates nothing if not the divine capacity for an excessive, abundant giving when formal reasons for hope are null. Rendered in the local Slavic language, Bosnian-Serbian-Croatian (BSC), "God willing" ("ako Bog da") is a popular saying among Bosnians of all ethnic backgrounds. Its near synonym in Arabic, inshallah, is especially favored by Bosnian Muslims whose devotional practice presumes, at the very least, a basic knowledge of key Qur'anic and ritual terms.

For Jusuf and Nijaz, who are not overtly pious Muslims, invocation of God is brief and low-key, and reveals what divinity means, promises, and delivers in minimalist vernacular references.

God alone was on his mind, Jusuf tells me, the many times when he felt his life was on the line. He tells of incidents during the Bosnian War and in the course of his deployment in Afghanistan when he silently, feverishly recited all he knew from the Qur'an. Ritual practice particularly unsettles some urban men of his age whose socialist Yugoslav upbringing presumed an opposition between a secular pop culture that was the public norm and the ethnic religious tradition that could only be lingering in the modern socialist worldview. Learned at home on occasions both mundane and particularly tender—as when parents and grandparents put children to bed with Qur'anic verses, uttered for blessings, protection, and just in case death surprised the darlings while asleep—Islamic mores up until the 1990s war were a private resource and a public embarrassment.

When the genocidal war broke up not just the socialist federation and economy but also the very ideal of a secular, ethnically neutral workers' fraternity underpinning the myth of Yugoslavia, Bosnian Muslim ethnic identity surged into realms of national and cultural politics, and with it, Islam regained a public face. In cities like Tuzla, Jusuf and Nijaz's hometown, which remained multiethnic in composition and socialist in political orientation, Islam becoming public presented to some Muslims a challenge to negotiate between the felt and personal faith and the social, performative practice of Islam. For his part, Jusuf joins the Friday tea hour at the old wooden mosque in the neighborhood, but he misses the communal prayers that had ritually gathered the rest of his neighbors as the faith community. He speaks of this with a laugh, describing the peculiar way in which he draws close to and defies participation in a Muslim collective.

Nijaz is similarly ambivalent. His wife wears a veil, which is as overt a sign of Muslim belonging in Bosnia as anywhere else, while their children's upbringing made Islamic rites a matter of second nature. The downturn of Nijaz's professional fortunes and his hurt pride are key elements in the story of his uncommitted position, neither secular nor reverent, in a family of prosperous, committed Muslims.

These biographical details flesh out the broader historical and cultural context within which faith utterances arise, always contingent on and colored by personal dispositions and outlooks—the way one inclines, wishes, or wavers with one's whole body. I am less interested in questions of piety—Muslim religiosity and its performance in a political context—than in understanding what Muslims' sayings, praying, and doings, their practices of reading and listening, propose about the nature of reality and the present state of the world.

This is not just a matter of grasping how people speak or experience reality but of understanding how a particular form of reality is engendered and at work between people and their God, whose names, after all, include the Real or the True (al-Ḥaqq). This may sound like I am speaking in terms of a symbolic-material "co-constitution," which is a favored concept in contemporary Western academia. The term suggests that meanings and matters are made through interactions of many agencies, which are imagined as equal parties "entangled" in horizontal relations. The interactions make the agents themselves through processes that are unfinished and underdetermined. What is here and now, is all there is but the eventfulness of the present in the making is imagined generously as a plenitude, an abundant immanence that could host anything, even unforeseeable futures. Even a god, provided that divine is one among many possible things, employed with "working out connections."[1] Contingency reigns supreme.

This processual idea of the world as always becoming and, contrary to appearances, unfixed and ultimately unbound by the formal powers that be, is a way to keep up hope and inspire progressive politics when they are needed most—at present. As the planet is divided by growing economic inequality and militant, conservative, or capitalist globalization, and rapidly devastated by environmental and climate catastrophes, keeping up hope is recommended with an urgency. It is to this end that the brilliant philosopher Isabelle Stengers recommends we embark on an "adventure of thinking," which is "an adventure of hope."[2]

Stengers argues that if we anticipate what is probable, there is no hope. Probability counts on a static world and the prescribed descriptions of it, while the world itself is in the constant process of becoming, with inventiveness and creativity manifest everywhere, in human language as well as in living cells. Adventure in thinking is open to wonder at this processual nature of the world and is primed by it, on the lookout for unscripted events that shift the way things are and feel. Possibilities cannot be calculated in advance, because they present themselves to the thinking unconstrained by givens. Hope itself, for Stengers, is a significant event, the kind that has the power to sustain wonder and gratitude, which in her writings are feelings and inclinations as necessary for scholarly and scientific enterprises as they are for social mobilizing for revolution or climate change action.[3]

I turn to Stengers, who calls herself a "daughter to the Western tradition" because her professed commitment to that tradition comes with an aspiration to resist the prevalent idea that the West has won the right to consider itself the "thinking head of humanity" and to presume, under the banner of critical thinking, "that others just believe and that we know."[4]

Keeping the philosopher's self-reflective note in mind, this sister thinker reads the Eastern and Western traditions while considering the implications of Islamic metaphysics for adventurous thinking about the world and for imagining and working out possibilities for hope, sweetness, and a brighter future against the odds. To that end, I ask, What do people who believe know about the nature of reality?

In the Islamic worldview, the cosmos is constantly in the process of being made anew. The God of the Qur'an, however, is not a deity that tolerates companions. Indeed, the greatest offense, *shirk*, is the association of partners with God, and so the reality the Qur'an describes is an utterly divine affair. And yet divine utmost competence presumes and engenders the possibility that humans, along with everything else (but unlike anything else)—the coming and going multitude that is the cosmos—exert a range of influences and efficacies. The One works through many causes but bluntly asks that the willful subjects make no mistake about the Doer who is with and integrates all the causes.

What seems like high theory has practical implications, for it matters where one directs prayers. *If you invoke them, they do not hear your supplication; and if they heard, they could not answer you*,⁵ the Qur'an says about misplacing trust in anything other than God. What also matters, practically speaking, is that the prayers are solicited with a strong recommendation: *Say, "What consideration would my Lord give you if it were not for your prayers?"*⁶ In other words, to say prayers is to catch God's attention. Beekeepers' hopes spring from faith in the reality that shows divine work through all events, such as honey flow—reality reworked at all times with an open invitation to court divine attention to one's circumstances in the troubling world.

Inshallah, God willing, is one of the most ubiquitous, mundane statements of faith. Its meaning varies in depth, depending on the speaker and the context, but its most casual utterance says something precautionary and buoyant about the baseline assumptions of how God works in the world.

In Jusuf and Nijaz's conversations, outcomes on the honey trail are regularly entrusted to God's will. "God willing" acknowledges the God who knows perfectly, the One who gives with good measure and who also withholds. Invocation acknowledges in advance the possibility of failure, as it asserts God's ultimate sovereignty over the lots drawn while it repositions humans and the weathered world as divine subjects. *And God does what He wills*, the Qur'an says.⁷ This is one of the formidable statements in the Qur'an that alternately describes God with qualities of nearness (tenderness or immanence) and distance (majesty or transcendence). It asserts that humans are not in control.

On a honey hunt, chances are wide open; even a windfall of manna could happen when the foraging season has drawn to a close. Hope makes sense not simply because the future is unknowable and utterly open ended but because the prayer itself is an active element in the world—tradition guarantees it—and because the source of all possibility to begin with is inclined to give goodness and to hear prayers. God's principal name is the Merciful (*al-Raḥmān*), and God wills mercy. As the Qur'an puts it: *God has prescribed for Himself mercy.*⁸

Beekeepers' hope bounces back through this basic assumption about divine will. On a deeper level, Islamic metaphysics premises the very possibility—the event of hopefulness arising within the mind or

Fig. 2.2 False indigo, a pollen riot

chest—on this divine self-prescription, because divine doing and willing encompass both the objective and the prospective, the visceral and the affective matters. In other words, feeling hopeful, like nectar flow, is a gift of divine mercy where depression and dearth could reign. Prayers actively summon hope—the Qur'an mentions a "breast expanding"—as well as nectar and clement elements.

## *Disappointments*

"Novice beekeepers are unlucky to be weathering another hard year," Šefik says, thinking back to 2017. "This past year, too, was a disaster. There are beekeepers in our association who are in their eighties. They've spent lifetimes in apiaries and have seen nothing like this. The frost in late April destroyed everything. But unevenly so. Here, on this hill, the forage was devastated, while there, one hill over, plants recovered and nectar flowed."

Šefik is president of the apiarists' association in the town of Tuzla to which Jusuf and Nijaz belong. In the late fall, this senior beekeeper shared his colleagues' impressions, which he has compiled from conversations and complaints of some hundred beekeepers of all scales.

Luck in beekeeping has always been weather-dependent and so highly contingent. Patience and perseverance paid off as the good-flow years made up for previous losses and costs. But patience is becoming another virtue, as bad years recur and the old rules of thumb no longer hold. Flow is spotty, disasters nearly total.

"Our expectations were high for black locust, they always are," Šefik goes on, reviewing nectar forage over the year, season by season. "Then you get so sad seeing it ruined. It blooms but does not smell. You know how black locust smells?" I nod at this knowingly: the crisp early scent of a year, petrichor, drenched with possibilities.

That black locust flowed at all was a wonder considering that the late April skies unexpectedly hurled snow and frost across the region. When frostbitten, black locust trees may not flower at all. Its bloom being timed to always shifty spring weather "makes the beekeepers' disappointment inevitable," says the *Atlas of Melliferous Plants*, an authoritative guidebook to regional honey forage.[9]

These are well-known facts, confirmed by years of experience, and yet they do not "inevitably" spell out a low horizon of beekeepers' honey anticipations. Quite on the contrary, high expectations, as Šefik makes clear, are the norm.

"It's sweet, intoxicating, the kind of smell that fills us with joy. Then, you watch it ravaged," Šefik adds wistfully. "Next, we look ahead, to linden, hopeful as if it were a saint coming."

At this, he smiles. He need not say more since, by then, I've learned well that linden is no saint; it is reputedly the most capricious nectar secretor. A trickster, it saturates the air with heavy perfume even while the nectaries are dry. Bees visit in scores. The insect pilgrims fall to the flowers loyally and, yielding little or nothing, keep returning for the length of the bloom.

Nodding to himself, Šefik concludes his reflections:

> From now on, considering weather conditions and global climate trends, we can only expect worse. We better get used to it. These are no longer exceptions, nor some extreme events, these are now the new conditions of beekeeping. A beekeeper is now in a situation where he has no gains, only expenses. It's tragic, this weather. It's the decisive factor. It used to be the

case that apicultural technology, the type of bee box and so on, made a difference, but weather is always the bottom line—it's God's giving (*Božije davanje*). If weather is bad, technology is of no use. Still, we have to keep trying, we have to search for ways out. Perhaps one way to do it is to diversify foraging sites; so if one thing fails there is another.

In Šefik's account, the evidence from the field and known facts about global climate change trends leave no room to doubt that apiculture is facing a catastrophic future. Because weather is a decisive factor to honey flow, extreme and unseasonal weather is already disastrous and the forecast for local beekeeping is nothing short of tragic. Even when the subject of global warming was rarely part of broader public discourse in BiH, many beekeepers' concerns were already stated in the language of climate change effects. Some actively followed international news on climate science, climate change diplomacy, and activism while sharing personal observations of unseasonal phenomena on social media with colleagues across the wider Balkan region and networking through social media. For professionals who were especially sensitive to the environmental toll of modern industrial culture, human responsibility for ruining the climate was obvious.

Apiculturists, however, belong to a community that prides itself on vocal disagreements. As the saying goes, two beekeepers can only agree on a single point, which is that you should not stand at the hive's entrance when opening a bee box. Understanding of climate change and its implications for local apiculture was uneven across the BiH beekeeping community, and younger beekeepers like Jusuf and Nijaz were often annoyed by what they perceived as their older colleagues' obliviousness to the signs of new and irreversible seasonal trends.

By 2017, Jusuf and Nijaz no longer needed arguments to convince their association's fellows, most of them their seniors in apicultural practice, that mobility was the future of beekeeping. It had become increasingly evident that stationary apiaries could no longer be expected to collect surpluses. In fact, hives on fixed locations could hardly be sustained without artificial feeding unless one extensively planted melliferous, honey-yielding crops, which, unsupported by irrigation infrastructure, were vulnerable to heat and drought damage.

In the quote above, Šefik, who was not a mobile beekeeper, admitted that the trusted apicultural techniques for improving chances of honey yields were proving inadequate and that traveling in search of diverse forage sites was the key strategy moving forward. Despite mounting signs of trouble and bleak future prospects, trying makes sense because weather, the bottom line of beekeeping, is God's giving.

Šefik is not preoccupied with sorting out the way tragic weather is an outcome of both human-caused actions and God's giving. He knows the basics: humans are held responsible for their actions, and highly contingent things like weather are stark reminders that everything comes through God's giving. Among the beekeepers and bee lovers are imams and Sufis with training and curiosity needed to delve deeper into matters. Contemplation is highly recommended by the Islamic tradition as an essential part of faith (as well as a form of pleasure) while, at the same time, the focus for metaphysicians and the curios folk alike remains on doing. The whole point of Islamic metaphysics is to teach people how to live better. Deep conundrums are not meant to be solved, given the nature of metaphysical concerns; they are resolved tentatively in the course of practice—doing and praying—which always emphasizes human responsibility for the acts, conducted and intended.

Šefik's unfinished contemplations on the divine and human agency in the making of the weather can be taken further by perusing cannonical Islamic sources. God in the Qur'an describes Himself as the Time. Some include *al-dahr*, the Time, among God's beautiful names. Several names describe God as eternal, while the basic tenet of Islam, that there is nothing but God—*lā ilāha illā Allah*—implies that the flickers of the present are parts of the everlasting reality, which is why any moment in the here and now births the possibility of encountering and getting to know God, through divine attributes, acts, and signs of presence. Moments of intensity and immersion, when we lose track of time, are events when the fleeting is eclipsed by the neverending. Moreover, the time being divine attribute, suggests that the generous God gives in time gifts that are as particular as the personal, embodied experience of time that we each have while the clocks' hands make the standard rounds. And God gives within history, each epoch manifesting a particular relationship between the world and al-Dahr.[10]

The final times in the Islamic conception of history since the last Revelation is not uniformly a time of loss, crisis, ruin, and anxiety but a long-lasting period marked by intensified possibilities. A number of

Prophet's sayings describe a rushed temporality whereby a year is like a month, a month feels like a week, a week flies by like a single day, and a day goes in a flash, "like kindling of a fire." Within time, however, there is the quickening of opportunities (to earn merit and profits, to gain understanding, and to earn wrath), a multiplication of access points to the divine as well as chances to miss them, and an overall inflation of values, fictitious and genuine. And because the final times presume the world running out of time, the bleak future moment when prayers, some hadith forecast, will no longer be accepted, the day when the sun rises from the west and the gates of repentance close forever, there is more need than ever to understand and invoke God.

Invocations rehearse what is known about God but also further one's understanding of God, since God is what exceeds one's conception, and divine generosity is promised but unpredictable. New circumstances raise new chances for invocation, and even if the utterances are simple and unchanged, their inflection is particular and the effort fresh, for the point all along is to engage God in a relationship that plugs the subject, God willing, into divine world making.

By 2020, itinerant beekeepers, too, admittedly struggled. Jusuf and Nijaz still offered classes on mobile beekeeping to the association's new members free of charge, but their apiary on wheels collected barely enough to make the traveling worthwhile.

### Gratitude

May 2021. Avdo, his two sons, and their helper are getting the hives ready. From a black locust forage site in the flatlands of the northwestern Bosna River basin, they will travel south. In a matter of some 300 kilometers (180 miles), the bees will traverse several climates, reaching the country's Mediterranean region. There, on the western banks of the jade-green, ice-cold Neretva River, in the valley stretched out beneath the steep slopes of the Prenj mountain range, the bees will roam the prickly bushes of Jerusalem thorn. Their yellow flowers, flat like shirt buttons, are just opening up.

Black locust, Avdo sums up, has partly flowered, and except for a few days, it flowed poorly. The calm, fit elder in his seventies, turns his intense blue eyes to the cell phone in his palm. He scrolls through the daily updates that the field electronic scale has sent as text messages over the past fifteen days. Fitted to one of the hives, the scale records weight gains, which helps the beekeepers discern shifting nectar trends.

"Last night, the inflow amounted to four hundred ten grams [fourteen ounces]."

"Is that little?" I ask.

"Of course, it's little! And it's a sign that the flow is at the end. I'll find you the days with better inflow, to compare," Avdo suggests, searching through the texts.

The beekeepers arrived with the first batch of bees on May 13, and the scale for a few days showed an inflow of up to 10 kilograms (22 pounds) a day. Then, in a week's time, on the twentieth, it dropped to a daily average of roughly 4 kilograms (8.8 pounds). At the end of a good flow, after the excess moisture evaporates, the hives are secured with 25–30 kilograms (55–66 pounds) of pure honey, which leaves plenty of surplus to harvest. As is, the honeybees have collected only enough to sustain themselves. Thank God, Avdo says, at least he did not have to feed them.

A few months earlier, in the Herzegovina valley where Avdo's bees overwinter, the spring was so untypically cold and the greening so late that the bees ate through the hefty winter stocks they had gained the previous year on the late fall bloomer: the Mediterranean white winter heather. "The bees were literally famished," Avdo says. He and his sons supplied the hives with emergency food; they gave them homemade pollen cakes.

He opens up a hive to show me how the combs look. I admire the few honey wreathes I see topping the combs, but Avdo finds them paltry: when black locust truly flows, the hives are clogged with honey from top to bottom, and the combs are quickly sealed with fresh wax, as white as snow. Through the 1990s, such flow was to be expected, but lately, it is rare. The last time it happened was in 2010, following years of dearth.

Avdo and his sons are among the greatest mobile beekeepers in the country. With some three hundred hives, the scale of their operation is impressive by local standards, as is the sheer range of their travels and the variety of honey flavors they chase. Avdo has been traveling with the bees since the early 1980s. Hailing from a family of apiarists, he began touring the honey routes of old Yugoslavia for fun. It was a welcome break from the small-town routines and his daily job in a secret weapons factory that produced ammunition for the Yugoslav Army. "We grew up thinking our father worked in a macaroni factory," his older son, Ado, tells us, laughing. The two boys were raised on honey itineraries. During school summer breaks, travels were the whole family's treat. The lavender-scented breeze on the blue islands in the Adriatic, grassy lakes in the central Bosnian mountain range to fish in, while bees foraged on thyme, and much more—the boys' days were sweet. They got the bug for beekeeping on the wheels.

The younger, Elvir, is the regional representative for a Swiss dental equipment company, but his flexible work schedule allows him time for the bees. The older, Ado, turned his back on a career in the food and beverage sector of the hospitality industry in Canada to return home to a full-time vocation in beekeeping. His income is incomparably lower, he says, but he is happy.

His sons' commitment helps keep Avdo going, although the beekeeping is becoming more difficult every year. "Everything has changed, believe me," Avdo tells me as we wait around for the foragers to settle in so the hives can be closed for tonight's trip. In the mellow southern accent of Herzegovina, Avdo shares his sense of the palpable changes in the familiar landscapes.

Black locust nowadays blooms roughly twenty days earlier than it used to. In 2010, the year of the most recent good harvest, the flowers opened on May 15. Lately, they bring hives as early as April 22 to meet the first bloomage. The later the tree blooms, Avdo explains, the greater the chances that it will flow uninterrupted by frost or sudden cold spells.

Except that "those rules are no more," Avdo adds. "See this year, such as it is." In 2021, spring was delayed across the country, and although black locust bloomed as late as it used to in the old days, the cold weather spell lasted throughout the tree's development and bloom. The overall effects, however, Avdo finds puzzling. Frost damage on black locust trees is usually evident: you notice flowers shriveled or blackened or the incipient stem stunted where the flowers would have been. But this year, such signs are absent, and he does not know what happened. Quite simply, roughly 40 percent of the trees did not bloom at all.

Aside from earlier signs of the spring, Avdo finds that the overall duration of nectar secreting has shortened. He remembers black locust flowing for a month back in 1996. The year after the Bosnian War ended, Avdo eagerly got the bees back on the road. Mountain meadow, too, lasted a whole month. These days, "no way it lasts more than seven or ten days." Plants' secretion, as Avdo puts it, is now "compressed." The forage outlooks are also far more fragmented. "Micro-locations now yield honey. Everything looks just the same but [some places] simply don't flow. That's that. Those are some of the changes."

Avdo's observations are culled from microforage locations, the smallest dots on district maps, within coves he revisits regularly across the forage fronts. But given the wide range of his forage pursuits, from the northern riverine lowlands across the central Bosnian mountains to the southern Mediterranean plains, his overall insights are regional in perspective. Moreover, Avdo's sense of diminishing nectar has a historical depth to it, emerging from forty years of traveling with bees.

"I don't know all that's going on but, for sure, it has do with climate. With pesticides, with our own doings . . . Some things I don't comprehend, but the changes are a fact. God help us," Avdo says. "Maybe that time [of good flow] will come again . . . but the chances are small. By God, I don't know . . . Some things I can't comprehend. A human does everything, tries hard, but it seems of no use. Still, I'm grateful. Had they stayed over in Herzegovina [at their wintering location], I would have had to emergency feed them."

*Remorse*

Avdo rarely sustains the tone of complaint for too long. While some mobile beekeepers are finding the daily accounting of hive scales too stressful to bear, Avdo has lovingly nicknamed the scale "Nura" after his small granddaughter, whose name in Arabic means "light."[11] That way, incoming texts from Nura make Avdo light up, even when the records reported are disappointingly low. He responds with expressions of gratitude. In a relationship with God and the world that belongs to God—*And to God belongs whatever is in the heavens and whatever is on the earth. And ever is God, of all things, encompassing*[12]—gratitude is essential for improving future prospects, but in a course of trial, turning to God begins with a *tawbah*.

*Tawbah* is a humble acknowledgment of one's fault. "It has to do ... with our own doings," Avdo said. It stands for "turning away" from doing wrong and for returning to God. A hadith says that trouble does not descend from the heavens except on account of people's wrongdoing and negligence, nor does it lift unless there is a *tawbah*. Islamic tradition has passed down articulate statements of *tawbah* as well as collective and personal rites that formally state a repentance. Communal prayers for rain in times of drought or for clear skies in times of storm, begin with a formal *tawbah*. The change in a weather forecast, however, is presumed to hinge on the change of human hearts.

Sufis in particular have emphasized the inner workings of *tawbah*. In the writings of Sufi and eleventh-century Islamic scholar al-Ghazālī, *tawbah* entails knowledge, a dawning realization of the harmfulness of a sinful or wrongful doing. Whatever the act and its outward implications, the harm has also been dealt inwardly in one's relationship with God. Doing what displeases God drops a veil between a human and the divine Beloved. The distance feels like a loss and overwhelms the heart. The feeling of longing stokes up a fresh desire and intent, to win back the divine Lover, by leaving the wrong at present and resolving to do better in the future. The remorse reflects on the past, too, on the wrongful act that one has both willed and, so, is responsible for, and that one has also been allotted, because divine will, ultimately, creates the choices that belong to humans.[13]

*Tawbah* is said to be the gift from God to the human species, among the bestowals that Adam and Ḥawwā,' the original couple, received in turn for the loss of paradise. A bestowal because the couple were first inspired to do a *tawbah* and then, having been forgiven by God, were given a promise that *tawbah* would be an aid in the course of human earthly life. "O Adam, I have entrusted you and yours with effort and striving and have entrusted you with *tawbah*, so whomever calls on Me, I will respond as I have responded to you, and whomever asks Me for forgiveness, I will not be stingy with it, because I am the Near, the One Who Responds. O Adam, I will raise those who perform *tawbah* from graves happy and smiling, and their supplication will be accepted."[14]

*Tawbah* prefaces supplication (*du'ā'* in Arabic, *dova* in BSC), which is an intimate counterpart to the formal daily rites. *Du'ā'* means "a call" or "a cry" and denotes a plea uttered under dire circumstances, when all reasonable grounds for hope seem lost, because God is not constrained by the given odds.[15] Forgiveness and rewards in the divine address related above are lavishly promised, and this bountiful divine readiness to respond is what recommends supplications, even in a casual style of soliciting God's protection and help, as when Avdo says "God help us." The Qur'an cites God with reassurances: *And when My subjects ask you about me—truly, I am near. I answer the call of the supplicant when he calls Me. So let them respond to Me and believe in Me, that they may be led aright.*[16] Also: *Call Me and ask Me, I will surely respond.*[17]

The Prophet taught his community that small, daily things are also worthy of prayers—ask God even for the salt in the house or for shoelaces, when needed, is among the hadith that popular collected volumes on supplications typically cite.[18] Nothing is too petty to concern God or too great to come forth from divine treasuries. Importantly, supplications are focused on the nature of divine generosity, not on one's merits or flaws. Al-Ghazālī writes: "Don't let what you know about yourself [your sins, your faults], prevent you from issuing a *du'ā*, for God has accepted the prayer of Iblīs [the devil], the worst among His creatures."[19]

A proper response to God's gift—and the prayer itself as an open form of personal address to God counts among divine presents—is gratitude and praise. Gratitude is a show of good manners and more than a precondition for a wish to be granted. A devout Muslim knows that gratitude is due under all circumstances. When circumstances are trying, a show of gratitude proves one's patience, which the Qur'an describes as a "beautiful" virtue.[20] When circumstances are fortunate, God, whose name is also the Ever Grateful (al-Shākūr), rewards gratitude with further gifts and blessings.[21]

Likewise, it is a show of good manners to not downplay the value of what one has received. If honey's harvest is modest, offer praise to God—*alḥamdulillāh*—who could give anything and gives wisely, such as it is. If you crave more, go ahead and ask, but snubbing what has been received is simply rude.

Ingratitude, the Qur'an repeatedly says, is a great offense and the trait of the faithless. The root word for faithlessness, *kufr,* implies the act of covering up or occluding. Both the ungrateful and the infidel, *kāfirūn*, conceal divine presents, as Sufi Shayk Ayne, my local guide to Qur'anic exegesis, tells me.

In a sourcebook on afflictions of the heart and ailments of the metaphysical human, which Shaykh Ayne's teacher Shaykh Mustafa Čolić has translated with a commentary from the sixteenth-century classic by Imam Birgivi, ingratitude is described as a "serious spiritual-heart illness."[22] It is the condition of taking things for granted, accepting things as one's rightful due rather than as divine gifts. Ungrateful are those who overlook the source of all giving and doubt the inherent wisdom of the measure and limit of each bestowal, whether the presents are sweet or hurtful. As everything falls under the purview of universal divine mercy and everyone subsists on divine care and provisioning, including the unbelievers, to exist is to enjoy divine gifts, by default. Gratitude, however, prevents the displeasure of the Gift Giver, while ingratitude earns wrath and eternal distance, for *surely, God does not love kāfirūn*.[23]

What is more, praise—in its customary form, "praise to God" or "praise to the God, the Lord [or Nurturer] of Worlds," or rendered in local language as *hvala Bogu,* thank God—joins the invoker with the praise that God gives Himself as His due. As is typically the case, elements of Islamic invocations and ritual practices promise benefits that are at once intangible and tangible in the down-to-earth sense. Gratitude is meant to cultivate a disposition and build character, deepen knowledge about divine doings in the world, and, ultimately, groom a faithful heart—a heart that has attained a divine quality, since God describes Himself as grateful. "Praise is the keynote of the blissful dwellers of the Heaven," as al-Ghazālī puts it in his discussion of patience and gratitude.[24] All things in the cosmos give God praise gladly, except for humans and jinn, who may miss or decline acknowledging divine blessings.

On the other hand, Avdo's quote "God help us" intimates the gravity of the present situation as he perceives it. The honey landscapes are waning irreversibly. Spreading wide his arms, Avdo shrugs his shoulders and asks: "What can we do?" He adds: "But keep trying hard."

Continuing to try hard entails clever scouting, swift movement, and ad hoc planning. The field hive scale's readings are helpful but by no means comprehensive guides on their own. Avdo and his sons still have to survey the forage area actively to get a sense of the particulars about

Fig. 2.3 Jerusalem thorn, on a windy day

nectar decline and arising forage possibilities. Sometimes it is enough to relocate hives short distances away to catch a better flow or to wait out bad days if there is a chance that the nectar will bounce back.

All things considered and reconnaissance done, it takes foresight to decide where to go next. At nightfall, as the men are loading up the hives into the back of the old baby-blue truck, Avdo tells me: "You know how it goes sometimes. We load up the bees, get going, and I still don't know where to take them next."

"Can you call someone up, ask for advice?" I wonder aloud.

Avdo laughs, his face glowing under the headband light: "All others are calling me for an advice! I decide, just like that, on the road."

His sons deeply respect Avdo and follow his lead. They do their best to spare his back the heavy lifting, though Avdo's energy seems to outlast everyone's, even when the three men are hard at work while keeping fast through the holy month of Ramadan.

## *God Knows*

In mid-June, my sister Azra and I meet Avdo again, by the hives in Herzegovina. Avdo is pleased: the bees have developed well in the last few weeks and are strong in numbers. The moderate flow of Jerusalem thorn so far is nothing to brag about, he says, but it keeps the hive reserves stable.

"What is his forecast for the rest of the Jerusalem thorn?" I ask.

"No one can tell you in advance. Who knows. Dear God alone knows. It'll flow, God willing," Avdo replies, firing off a laugh.

Because no one can tell the future of flow, unknowability is the baseline for nectar prospecting. The invocation of "dear God," however, at once sharpens and softens the edge of uncertainty. The flow, like all things, rests within divine knowledge, which encompasses all things. Several lines from a well-known Qur'anic verse, known as the verse of the Throne, which Muslims around the world learn by heart, read: *He knows what lies before them and what is after them and they encompass nothing of His knowledge save by His leave. His throne includes the heavens and the earth, and He never tires of preserving them.*[25]

In the writings of Ibn al-'Arabī, a thirteenth-century Sufi and an everlasting inspiration for Muslim thinkers, the "Throne" refers to God's knowledge, and God's mercy encompasses everything, including God's knowledge. God wills mercy to be the primary quality in his relationship with the world, and since mercy is all-encompassing—limitless—there is nothing in the heavens or the earth other than mercy.

The greatest show of mercy is the gift of existence itself, which is constantly replenished. Al-'Arabī reads, in translation: "He is the One who never repeats a single thing in existence. Possible existence, that is entities, have no end. New instances are engendered every moment without repetitions and without retaining the old form and that process is everlasting in this and in the next world."[26] The possibilities are unlimited, what comes through to particular beings depends on their constitution, states, and circumstances, all of which are subject to change. Given its source—mercy and knowledge—all giving is good, though the subjects may not like the taste of what they receive.

It is God who *makes you laugh and makes you weep*,[27] the Qur'an says, recommending composure and gratitude on the occasion of losses as well as gains. Trust in God's knowledge ought to put things into perspective, *so that you may not despair over what escapes you nor get overjoyed over what you have been given*,[28] as the Qur'an cites God. Mentions of God's knowledge, in the Qur'an and in Muslim conversations, are meant to humble with reminders of the inherent limits to human understanding and the short-sightedness of human desires. The following verses are often restated for comfort and encouragement: *You may hate a thing though it is good for you and you may love a thing though it is bad for you. And Allah knows while you do not.*[29]

Avdo remembers the days when Jerusalem thorn yielded up to thirty kilograms per hive, but lately, the most they get is five kilograms per season. If they are lucky, the hives will add more honey to the black locust stocks still sitting in the combs. Each subsequent forage may add to the mix, and by the end of the summer, they hope to have enough to harvest.

On the other hand, God is known to be nothing if not generous. Faith in divine generosity is a matter of principle, which ritual practice restates daily—at the very least, through the customary invocation of God as the Compassionate, the Merciful. Copious divine giving, moreover, is a matter of fact: it is on the honey records.

"Of course, if it starts flowing well here, we'll harvest before moving," Avdo says with glee. Up until the last minute, the flow can surprise them. Sometimes, beekeepers come to load up the hives only to find them suddenly frantic with foragers collecting nectar. A dull, dry landscape can turn honeyed overnight.

One year, Avdo remembers vividly, something flowed here like never before. Within days, each hive gained thirty kilograms of dark, sparkling honey. They never discovered the nectar's source. The thickets were alive with strong chirping, which convinced Avdo that it was the frosted-moth bug's that excreted honeydew. The frosted-moth bug had not been spotted in the area previously, nor ever since. When Jerusalem thorn stops flowing, the bees will be off to the mountains of eastern and central Bosnia, searching for meadows and forest honeys.

### *Chasing after Summer Heather*

Catching up with husband and wife Sabaheta and Nedžad is never easy. Seasoned mobile beekeepers, the couple learned the tricks of the trade from Nedžad's parents, who toured the former Yugoslavia with a bee caravan. When the old man retired, mourning the loss of his dear wife and road companion, the couple inherited the truck and the routes.

Those honey routes shrank after the breakup of Yugoslavia to only those territories within Bosnian borders. Emerging from years of siege in a central Bosnian town, the couple took the bees back on the road when the war ended. Beekeeping, however, was no longer the same.

A voracious new pest, the varroa mite, which was found to sap apian lymph and induce secondary infections, was now a widespread threat to the hives' health. Monitoring mite infestation levels and treatments became an important, all-season routine. Local environments and climates have changed, and, in response, the couple's trips are becoming shorter and more intense and their forage itineraries more changeable. The feat of staying on the honey's trail with nearly three hundred hives is beginning to wear down the beekeepers as they age. Their two children grew up with the bees on the road, but the adult son and daughter do not care much for apiary affairs.

By early July, once black locust, indigo bush, and linden forage have passed in the northeastern lowlands and chestnut blooms are spent in the country's north or west, the couple takes their bees to the Kupres highlands in the country's southwest.

At over one thousand meters (thirty-three hundred feet) above sea level, vast valleys spread out between evergreen mountains, their slopes graced with small emerald-green lakes. With the local population steadily emigrating, once-cultivated plains are left untilled while the short growing season attracts the country's itinerant shepherds and beekeepers to the extraordinarily diverse meadow plants.

Highland meadows typically offer a steady, long-lasting flow. Pine trees in the area are known to drip with manna, or honeydew. The couple park their hives on the land of their local hosts, a Catholic family of Bosnian Croats, Mario and Snježa. When they first met, the couple was unemployed, struggling to support their four school-age children. Within a few years, the beekeepers have trained them in apiculture, and, with a gift of several hives, the couple has built an apiary of nearly forty hives. Sabaheta also helped Snježa start a line of natural cosmetics by showing her how to cultivate and collect medicinal herbs and render them into facial creams and lip balms. Mario's mother fondly calls Nedžad "a saint" and kisses his cheek. But the sensitive subject of their war histories is carefully broached, just like their social encounters deliberately curate their ethnic and religious differences. They barbeque together in the summers, but their plates and glasses, intentions, and bellies mark signs of significant dietary distinctions: pork and wine do not cross the table from the Catholic to the Muslim diners.

As the meadow mellows, the beekeepers start scouting the lower altitudes for the main attraction of the region: summer savory, or *vrisak*, as it is locally known. Growing farther south, in the more humid and temperate climate at seven hundred meters, bushes of summer savory, with perky, bladelike leaves, bloom in a flash.[30]

Years of chasing after vrisak have given Sabaheta and Neždad a thorough feel of the valley, but this memory map cannot plot their itinerary in advance since this type of vrisak never blossoms twice in the same spot. Whether and where vrisak will flow is the question they take to the road each season. Following the trail of humidity and the shifting directions of wind and rainfall, they search out promising sites. The fugitive flower is well worth seeking since it promises a flow that can theoretically last from July to October, or at least a month, which is a lot. The couple cover considerable distance in their search on a two-lane highway, past the 1990s war remains and shells of blasted houses that have been overtaken by trees, brambles, and plants. A few homes have been fixed up by returnees, their livelihoods broadcasted by the road signs advertising sheep and goat cheese.

To find the flowers, however, Sabaheta and Neždad slowly maneuver the truck off the road and descend across sharp highland grass and much stone into the blooming fields. They cook and sleep by the offloaded truck, sometimes in a trailer. Sabaheta's back suffers in the field. Neždad gets burned out. If it turns out that the vrisak on the spot is not flowing, they load up the hives and search onward. Not too long ago, they drove 2,000 kilometers (roughly 1,240 miles) up and down the same road, searching for a spot that flowed.

The last good year for vrisak was 2014. The couple remembers that the scale recorded a daily income of 5 kilograms (11 pounds) of nectar. Since then, daily yields maxed out at 0.5 kilograms (1.1 pounds). Melliferous forage, in general, is less yielding, the two beekeepers say.

"The climate has simply changed, everywhere. Sometimes everything in nature looks all right, but nothing comes about. We cannot influence that," Sabaheta says. Neždad is the quiet type, so she does most of the explaining. What the beekeepers can manage is their own mobility. They expand the range of their travels. Whenever possible, they spread hives across several locations within the same season. While half of their hives are in Kupres, for instance, other hives are ready to meet the meadow flow at the Vranica mountain range farther north.

Neždad monitors daily electronic hive scale reports and keeps scouting the areas of potential flow. Fellow beekeepers keep calling with inquiries and cross-country updates on what is flowing and what has dried up. These conversations range from guarded to frank, among friends. The countrywide perspective helps everyone get a better sense of the year's trends as well as to decide how and when to make and shift plans. Still, nectar-flowing spots are professional secrets, and if the latest news is indiscriminately leaked, scouting can be spoiled by too many beekeepers on the same trail.

### Say "God Willing"

With bees in a fixed location, Sabaheta says, you have to be lucky to get a harvest, and even so, forage opportunities are limited to what grows nearby. "On the wheels, you have to keep going. If you want a variety of honey types and if you want to keep the bees at all, you have to search for the flow, to come to the bees' rescue." Whereas in the past they would find a good spot and could camp out by the hives for much of the plants' blooming season, nowadays, they move quickly at the first signs of nectar oscillation.

"We cannot really plan anything, nature dictates the terms. You're looking at the sky, sorting out the signs, watching the weather, which determines everything. Then other things can happen. Sometimes, something comes down with the rain. I have seen vrisak dry up after showers. Just like back home, one year, rain fell on tomatoes and the next day they turned black. Black!"

Vrisak is a mysterious plant to begin with. "One never knows what it takes" for its flowers to well up with sweetness, Sabaheta says, for "it's a strange plant! Strange." Conditions may seem perfect: moderate air temperature with the right amount of rainfall, and yet the flowers may not get "activated," as she puts it. Lately, the elements have also become weird: the winds unseen before, poisonous rains.

But neither the weird new weather nor the textbook-case atmospheric conditions for secretion ultimately determine the particular flow. The regional guide to honey hunting, now a dated sourcebook from the 1980s, describes decisive conditions for nectar flow of two species of vrisak: plenty of dew at night with morning and noon air temperatures in the range between 12°C and 32°C (54°F–90°F ), respectively. Sabaheta and Nedžad, however, know from experience that flowers can secrete against all odds. They remember a year when a dramatic *bura*, the wind picking up speed on its course to Bosnia from the Adriatic Sea, did not disturb the nectar flow. Another time, vrisak flowed through unseasonably low temperatures: frost-covered blooming fields attracted bees as soon as the morning air warmed. "As ever, dear God gives when you least expect it, even when conditions seem bad," Sabaheta says.

In early August 2017, at the turn of vrisak season, I call Sabaheta for the flow forecast. "We shall see what awaits ahead," she says. "We are all in God's hands, and who knows what may be. We are now talking but who knows what can happen even as we hang up. The time has come that you cannot plan, not even simple things, like 'I'll meet a girlfriend for a coffee.' Say 'inshallah' or 'ako Bog da.'"

"You know the story about the man who said 'I'll buy a cow today'?" she asks me next. I do not, so she goes on to tell it.

"'Say inshallah,' someone advised him. 'What for? It's as good as done,' he replied. 'The money's in my pocket, and it's a market day: I'll buy a cow!' So to the market he went, found the right cow, reddish, gentle, with a promising udder, reached for the pocket and found it empty. 'Inshallah,' he said, 'the money's gone.'"

We share a laugh and on that buoyant note finish a conversation that all along rang with undertones of the beekeepers' road fatigue and latent anxiety about what would happen next.

"God willing" in this story emphasizes caution about statements of intent. Two verses in the Qur'an advise so explicitly: *And never say of anything, "Surely, I will do it tomorrow" without adding "If God wills."* [31] Sabaheta's wisdom story, in fact, retells in a light tone a cautionary Qur'anic tale about owners of a garden who vowed to harvest its fruit without making an exception for God's will. The next morning, they found the garden crops ruined.[32]

Invoking God's will redraws, on the go, the map of the cosmos in which no doer is a god: neither the honeybees nor summer savory, not the weather, not auspicious sites, not even determined beekeepers' field techniques decide the outcomes of their singular encounters. Possibilities are endless, vagaries of weather notorious, and the current nectar trends discouraging. Saying "God willing," however, also pleads with God, who recommends leaning on Him and who replenishes all possibilities, including the possibility of hoping and praying when formal reasons may suggest giving up effort and when frustrations may stoke doubts. A statement of confidence in God—who promises to respond to calls—renews faith, not simply as a willful belief ("blind faith," as the saying goes) or a product of a formative experience in an object-subject encounter but as the changing receptivity of the one who is faithful.

The heart of a believer is said to be between two divine fingers, always turning between hope and fear. Invocation is meant not just to direct God's attention to the atmosphere, to flower nectaries, and foragers' routes and to tilt chances in one's favor but also to enhance one's receptivity to divine gifts, including the gifts of hope, patience, and gratitude, through honey flow or dearth, whether it is through laughter or tears.

## *Practical Metaphysics*

Prayer in the Islamic tradition comes with a fabulous promise, but it is a curious medium insofar as it poses a metaphysical problem. Namely, it purports to move the ultimate Mover, to compel God, who, the Qur'an explicitly says, does what He wants and knows in advance what comes forth. The idea of God being swayed by something contingent—a human or anything else in the cosmos—is therefore preposterous on one level of consideration. On another level, it is limitless divine mercy that promises God's responsiveness to the point where prayers are said to be capable of changing one's destiny. Nothing less than that. This tension between human and divine agency, between divine imminence and transcendence is never resolved, but the bottom line is always in favor of the practice, which, tradition says, offers ready advice: do pray, since prayers will be heard, as the Qur'an says.

Metaphysicians are bent on practice. While writing on repentance and gratitude, al-Ghazālī at one point ponders the problem of human choice in doing and wrongdoing. He takes up several pages rehearsing theological arguments of all stripes, ranging from completely free human will to absolute divine predetermination to the middle grounds staked between the two positions. He thinks through causes and effects and considers forms of insight that integrate positions that seem contradictory, only to leave the question unsettled. Instead, he says, "But let us go back to our business."[33] And his business is to recommend to the readers to practice repentance and gratitude and to elaborate on their meanings and merits. The whole business of practical metaphysics is to encourage contemplation while explaining how to better practice faith.

This is true especially in the times of trouble. The local beekeepers' prayers are uttered from within landscapes and atmospheres that are rapidly changing and, they anticipate, are becoming inauspicious for the honeybees. Earlier springs and longer-lasting seasons due to climate change, pollination ecologists suspect, could potentially disturb plant-pollinator seasonal matchmaking and even entirely uncouple plants and their favorite insects. Most studies to date have focused precisely on this anticipated effect of climate change. There are not many studies, however, and those that do exist are limited to relatively few species of plant and insect partners. The findings so far are also inconclusive, with evidence of both mismatches and synchronous adjustments. The effects on honeybees are underinvestigated. Bosnian Muslim beekeepers offer firsthand observations on how shifts in seasons and changeable weather disjoin honeybees and the plants that used to provide them with nectar and pollen. Apiculturists are managing recurrent mismatches with artificial food, which, as the next chapter shows, is not so much a solution as a short-term strategy for managing the crisis.

Whereas climate biologists are concerned about the trend of springtime advancing in the northern hemisphere, local apiculturists anxiously point out that the climate conditions throughout the year no longer uphold the customary distinctions between the four seasons. Put simply, season as a category now barely holds. In addition, and this may be the most precious insight from the field, apiarists are reporting qualitative changes to the plants, what biologists might recognize as alterations in plant traits and physiology, alterations with consequences for floral scent, for the quantity of bloomage, as well as for the abundance and composition of nectar and pollen. Climate biologists have acknowledged a dearth of studies on this subject. Long-term observations of climate-related changes in phenologies, plant physiologies, and plant-pollinator interactions more generally are greatly needed but time-consuming and costly to undertake.[34] Climate biologists are certain, however, that the responses of living organisms are going to be highly idiosyncratic. The world of living difference can be expected to respond in myriad ways, which makes the future of plant-insect relations highly unpredictable.[35]

Nectar flow is becoming fickler than ever. As troubling as this trend is, the events of honey's waning and surprise yields accentuate the work of generosity, as a divine attribute that is unforeseeable but also pledged to those inclined to courting it.

### *Honey Harvest*

My sister Azra and I visit Avdo and his family near the end of the honey season in 2021. The great round table on their terrace, in the shade of ripening fig trees, is spread with honeys. We are tasting the flavors of this year's harvest. Avdo's granddaughters Nura and Uma, ages seven and four, are also having a taste.

"This year was terrible," Avdo says. To my sister Azra and me, who barely managed to harvest twenty jars of linden honey from twenty hives, the range of flavors displayed on the table seems impressive. There are seven jars in seven different hues, each enclosing a bouquet of aromas: black locust honey, the color of parchment; golden linden; meadow honeys in shades of umber reds; and the obsidian-dark forest honeydew. Avdo, however, shakes his head doubtfully: "This is thin. Considering the number of hives and sites visited . . ." His older son, Ado, adds: "But, *alḥamdulillāh*. We've managed to gather something."

"Our *babo* (father) often says," Ado carries on, "the way things are going—and if we live long enough we'll know for sure—a jar of natural honey will be a miracle in ten years' time. I don't know what's going on, but there's less of it every year. May I be mistaken, oh my Lord."

"Amin," we join in.

## Notes

1. Adam Miller, *Speculative Grace: Bruno Latour and Object-Oriented Theology* (New York: Fordham University Press, 2013), 1–3. An object-oriented-philosopher Adam Miller adapts Christian epistles to develop a thingly theology.
2. Mary Zournazi and Isabelle Stengers, "A Cosmopolitics—Risk, Hope, Change," in *Hope: New Philosophies for Change*, ed. Mary Zournazi (New York: Routledge, 2003), 245.
3. See also Isabelle Stengers, *In Catastrophic Times: Resisting the Coming Barbarism*, trans. Andrew Goffey (London: Open Humanities Press with Meson Press, 2015) and "Reclaiming Animism," *e-flux journal* 36 (July 2012): 1–12. https://www.e-flux.com/journal/36/61245/reclaiming-animism/; Stengers, "Wondering about Materialism," in *The Speculative Turn: Continental Materialism and Realism*, ed. Levi Bryant, Nick Srnicek, and Graham Harman (Melbourne: re.press, 2011), 368–380; Stengers, *The Invention of Modern Science*, trans. Daniel W. Smith (Minneapolis: University of Minnesota Press, 2000).
4. Zournazi and Stengers, "Cosmopolitics," 250. Stengers has attempted to think alongside neopagan witches, Virgin Mary pilgrims, Quakers, as well as physicists and Western philosophers. See Isabelle Stengers, "Including Nonhumans in Political Theory: Opening Pandora's Box?" in *Political Matter: Technoscience, Democracy, and Public Life*, ed. Bruce Braun and Sarah J. Whatmore (Minneapolis: University of Minnesota Press, 2010), 3–34.
5. 35:14.
6. 25:77.
7. 14:27.
8. 6:12.
9. Veroljub Umeljić, *U svetu cveća i pčela: Atlas medonosnog bilja 1* (Kragujevac: Veroljub Umeljić, 2006), 199.
10. See Peter Coates, *Ibn 'Arabi and Modern Thought. The History of Taking Metaphysics Seriously* (Oxford: Anqa Publishing, 2011).
11. A mobile beekeeper from central Bosnia, Mehmed thinks that scales are becoming less popular among his colleagues. He retired the field scale he owned. "These tools are too stressful," he says, laughing. "Those who have them, stress out every day when the text [report] arrives. The rest of us stress only on those days when we visit the hives, have a look, and give them a tug. I find it easier this way. I live in hope, for three or four days and then, on the fifth day, if I'm crushed by the findings, I'll manage to fall asleep somehow." Mehmed hardly seems the type to stress easily. A composed man in his early fifties with a bright disposition, he is an imam by training. Since he resigned from his post at the local mosque, he has invested all his energies into the apiary. A resourceful keeper, he says flatly that he is determined to stick with the bees, no matter the losses. The precise microrecords of nectar's oscillation, however, are unnerving. By contrast, the suspense helps nourish upbeat expectations.
12. 4:126.
13. Abu Hamid Muhammad Al-Ghazālī, *Ihja' Ulūmid-dīn, Preporod Islamskih Nauka, 7*, trans. Salih Čolaković (Sarajevo: Libris, 2010), 7–20.
14. Ibid., 16–17. I translated this section from al-Ghazālī's pages, but the story of *tawbah's* bestowal, as is so often the case with important lessons on Islamic conduct, is a part of oral Bosnian Muslim lore. Shaykh Ayne often retold it in versions that improvised on al-Ghazālī, his favorite author, Qur'anic verses on tawbah, and stories he learned from his father, an imam, as well as from Shayk's own Sufi elders.
15. A *du'ā'* is not necessarily unscripted. Favorite prayers are passed from the Prophet of Islam, his family and friends, other prophets, and many of God's friends, great Sufi sages among them. They are published in collections and made into popular booklets, passed by word of mouth and social media, hand-copied or photocopied, learned by heart, or read out loud, regularly or in case of need.
16. 2:186.

17. 40:60.
18. See Jusuf Tavasli, *Dove i njihovi fadileti* (Sarajevo: Libris, 2013), 25.
19. Abu Hamid Muhammad al-Ghazālī, *Ihja' Ulūmid-dīn, Preporod Islamskih Nauka*, 2, trans. Salih Čolaković (Sarajevo: Libris, 2010), 322. On the devil's prayer, see chapter 4 of this book.
20. 12:18.
21. 14:7; 3:145.
22. Mustafa Čolić, *Et-Tarikatul-Muhammedijjetul-Islamijjetu, učenje i moral Allahovog Poslanika Muhammeda a.s.: Srčano zdravlje i bolesti metafizičkog insana* (Visoko: Tekija Šejh Husejn baba Zukić, 2016), 473.
23. 3:32; Čolić, *Et-Tarikatul-Muhammedijjetul-Islamijjetu, učenje i moral Allahovog Poslanika Muhammeda a.s.*, 474.
24. Al-Ghazālī, *Ihja'ulūmid-dīn*, 7, 208.
25. Korkut, *Kur'an s prijevodom*, 2:255.
26. Muhyiddin Ibn al- 'Arabī, *Mekanska otkrovenja*, trans. Salih Ibrišević and Ismail Ahmetagić (Sarajevo: Ibn Arebi, 2007), 458–460.
27. 53:43.
28. 57:23.
29. 2:216.
30. There is some confusion over this plant. Vrisak and *vrijesak* are the same name rendered in two dialects, and Mrka sticks to the southern inflection: vrisak to distinguish it from the kindred plant growing in the northeast: *Calluna vulgaris*, common heather. But two types of *Saturea*, summer and winter savory, are more difficult to sort out, and the dialectical difference is of no help. Mrka identifies the plants they seek as summer savory, *Saturea subspicata*, while an authoritative guide to the melliferous flora in the region, compiled by migratory beekeeper and author Veroljub Umeljić, identifies the blue blossoms as *Saturea montana*, winter savory, and the white blossoms as the Mediterranean plant. Veroljub Umeljić, *U svetu cveća i pčela: Atlas medonosnog bilja 1* (Kragujevac: Veroljub Umeljić, 2006), 664. White and blue blossoming vrisak alternate in the area of Mrka and Nedžad's scouting, although the blue vrisak is by far more prevalent. In this text, I stick to vrisak.
31. 18:24.
32. 68:17–18.
33. Abu Hamid Muhammad Al-Ghazālī, *Ihja' Ulūmid-dīn, Preporod Islamskih Nauka*, 7, trans. Salih Čolaković (Sarajevo: Libris, 2010), 21.
34. See Joar Stein Hegland, Anders Nielsen, Amparo Lazaro, Ann-Line Bjerkens, and Orjan Totland. "How Does Climate Warming Affect Plant-Pollinator Interactions?" *Ecology Letters* 12, no. 2 (February 2009): 184–195, https://doi.org/10.1111/j.1461-0248.2008.01269. See also chapter 4, note 40.
35. See Simon G. Potts, Jacobus C. Biesmeijer, Claire Kremen, Peter Neumann, Oliver Schweiger, and William E. Kunin. "Global Pollinator Declines: Trends, Impacts and Drivers," *Trends in Ecology & Evolution* 25, no. 6 (February 2010): 345–353, https://doi.org/10.1016/j.tree.2010.01.007; Eric Post and Michael Avery, "Phenological Dynamics in Pollinator-Plant Associations Related to Climate Change," in *Biodiversity and Climate Change: Transforming the Biosphere*, ed. Thomas E. Lovejoy and Lee Hannah (New Haven, CT: Yale University Press, 2019), 42–54; Camille Parmesan, "Range and Abundance Changes" in *Biodiversity and Climate Change: Transforming the Biosphere*, ed. Thomas E. Lovejoy and Lee Hannah (New Haven, CT: Yale University Press, 2019), 25–38. On considerations of extreme climate weather events and more general uncertainties surrounding future biodiversity conservation policies see of Pablo Marquet, Janeth Lessmann, and M. Rebecca Shaw, "Protected-Area Management and Climate Change," *Biodiversity and Climate Change: Transforming the Biosphere*, ed. Thomas E. Lovejoy, Lee Hannah, and Edward O. Wilson (New Haven, CT: Yale University Press, 2019) 283–293.

# – 3 –

# Planting on the Eve of Apocalypse
## The Prophet's Advice

Imagine you set out to plant a tree one day. A sapling is ready in your hand. Its trunk is smooth and slender, its riotous roots eager to dig in. The tender leaves foreshadow fruit, flowers, and shade, a dwelling for shy birds and ravenous insects. You get to work.

The soil yields to the shovel's edgy advances, when suddenly the ground violently shakes. Your gut somersaults. The daylight sours, *the sky is as molten lead*.[1] Hills uproot and fly by as if they were tumbleweeds swept up by winds, unspooling along the way *like carded wool*.[2] You stumble, your feet as wobbly as rubber. The sun rises from the west. What to do?

Go ahead. Plant that tree, the story says.

I am retelling a story that Bosnian Muslims are fond of telling each other. Each teller relates the details somewhat differently, but the eve of the apocalypse they describe conveys Qur'anic imagery or, at the least, a Qur'anic sense of the Hour when the earth is beyond rescue. In some retellings, you are advised to plant on the eve of your death. The source of the story is a hadith, a saying by the Prophet of Islam, and its retelling passes on prophetic warning along with good advice. I have heard the story told to recommend planting but also to encourage doing and intending good, even when one's efforts seem futile or thwarted by dire circumstances. The upbeat message seems obvious enough: one should keep trying for as long as one lives, even if the skies are falling.

Still, it is worth asking the obvious question: Why plant on a doomed planet? Whereas ecological activism typically springs with hopes of preventing or deferring a catastrophe, or at the very least counts on the possibility of remediating its disastrous effect, this story takes us over the edge, where the time is up and nothing can be done to prevent the world's collapse. And yet, do plant, is the prophetic advice.

The wisdom that the hadith and its popular retelling convey is eco-eschatological in the sense that the ecologically mindful sentiments and actions it encourages are decidedly antiapocalyptic, and yet their meaning derives from the precarious state of the world that, precious and finite, is inevitably going to ruin.

This chapter relates three Muslim beekeepers' earnest planting efforts on the grounds of their apiaries and beyond while pondering this popular story and prophetic wisdom in the wider context of Islamic ecological sensibilities at work in planting and doing service in the world.

Gardening is popular in Bosnia. Gardens and orchards have traditionally been integral to the local subsistence economy, and their importance has only grown in recent years. Their bounty stocks local kitchens and pantries with fresh produce, and grains, beans, and frozen and preserved fruit make up winter diets. The economic value of homegrown produce is significant for the average citizens, who live on relatively low incomes, and for the many unemployed Bosnians.

Local gardening, however, is more than a bare necessity. Growing one's own food and collecting fruit have been part of a more general effort to secure nutritional and remedial diets in the national food market, where environmentally friendly farming and eco-labeling are not formally established but anxieties about food contamination by agricultural and industrial chemicals are high. Ideally, gardens are situated far from sources of urban pollution, industrial contaminants, and agricultural chemicals, but just as often, serious planting is underway in the midst or along the edges of extremely toxic landscapes.

Beekeepers tend to be especially enthusiastic gardeners, avidly planting at their apiaries and beyond, across the lands untended by their absentee owners. Given the high regard for the nutritional and medicinal attributes of honey and its value for both human and apian health, local beekeepers have traditionally planted to secure environmentally clean forage for their bees.

They have also cultivated pollen-rich and nectar-yielding trees and plants to sustain honeybees through dearth and to improve honey harvests. Planting for the bees' sake seems more necessary than ever, just as it is proving ever more inadequate to counter the felt effects of climate change on honeybee ecologies. This is why beekeepers, of all local gardeners, are the best companions with whom to ponder the feat of planting on the eve of apocalypse.

## A Hunch: Observing the Signs

"I have a hunch," Šefik says, "a sign [*išaret*] of sorts, though it's only what we think, but everything will change." He is off-loading sacks stuffed with bottles of sugar syrup from the back of his van into a wheelbarrow. Senada, his wife and work partner, is unlocking the gates of their apiary. We are parked at the end of a dirt road, at some distance from a small village in Western Bosnia. Nearby creeks flush the air and meadows with cool morning dew even throughout the summer heat waves of 2021. The grass is so wet that the couple, taught by experience, are wearing tall rubber boots. My sister and I, who came to film the beekeepers at work, are bound to get our sneakers miserably wet.

"In ten years' time, everything will change, and our lives will be entirely different... Just like ten years ago, I used to say that honey will soon be packed in very small jars." That time has already come, Šefik says, slamming closed the back door of the van.

"All that's happening in nature and on our planet Earth, I believe, are God's decrees, and I believe the human has caused all this trouble." Pausing to stretch his back, he spreads his arms: "See this, all that's surrounding us." In the silence that follows, our ambience comes forth to meet the senses: deeply green, serene, smelling of damp moss and Caesar's mushrooms hatching throughout the woods. The land surrounding the apiary is flanked on one side by old oak, linden, and spruce forests, their crowns reaching dark and grand far above us. The other side looks out onto open plains, unfolding with meadows as far eastward as the eye can see, where the land and the sky, unawake, lie as one.

"It's a God's gift. And we? We need to give every day, too, to be happy [*sretni*]. We can talk more about that," Šefik says, laughing, and pushes the barrow through the garden gate and into the apiary, its wheel squeaking its lone tune amid the quiet hives.

Nearly one hundred hives, painted in yellow and blue, are spread out in a densely planted bee yard. Šefik lifts the lid off the last hive in the first row, resting beneath the canopy of a handsome fig tree. Meanwhile, Senada, a quiet, wise woman, down-to-earth and caring, works the hives in the next row.

"This is cruelty to her," Šefik says, pouring sugar syrup from a recycled Coke bottle into the hive feeder. "Her" stands for the bee, commonly referred to in the local language by female personal pronouns. Šefik speaks in a raspy morning voice, his figure a mere silhouette against the predawn horizon.

"This food is bad for her, but we have to give it when there's no food in nature. We can talk about why we do that, but the bottom line is that honeybees can no longer survive without human help. At least not around here, where we live. We didn't used to do it, but we are now struggling to keep the bees alive. And all we're doing now, all year long, is struggling to keep the bees alive."

It is August 2021, and Šefik and Senada are feeding the bees every second day to help them build winter food supplies. They feed them early in the morning to prevent hive robbery—the incidents of food raids committed by other honeybees, highly likely under the conditions of dearth. They feed them consistently while the population of summer bees is still active in the hives. Processing the sugar syrup to build honey stocks, local beekeepers feel, is strenuous work, which the winter bees should be spared. Due to hatch in these hives within weeks, winter bees, ideally, live semidormant lives, preserving their energy and bodily fat reserves until springtime when they help build back the apian population for the new foraging season. These bees' bodies will produce the vital substances that nurse the egg-laying queen and help raise the new brood.

The local beekeepers consider sugar syrup a poor substitute for honey, which is praised for its nutritional, prophylactic, probiotic, and therapeutic properties.[3] Sugar and pollen alternatives, commercial and homemade, however, have become indispensable not only to supplement the hive's low honey reserves at the close or start of the foraging year but also to come to the bees' rescue during the recurrent and long-lasting episodes of dearth. A bland sugar diet shortens the bees' life span and compromises their strength and immunity.

"Instead of living for forty days, as they used to, bees fed on sugar can die within fifteen days. And so, what happens? With artificial food, we manage to produce the bees in numbers, but they die too quickly because they lack that goodness from nature, which is the God's gift."[4] "Luckily," Šefik adds, "the bees here manage to collect plenty of pollen; otherwise, they would perish."

"What about the nectar?" I ask.

"Look, we are now beekeeping under altered agro-meteorological conditions," Šefik says. By now, we have reached the apiary's third row. The sunrise lights up the tin hive roofs with a rosy glow.

"The climate has changed and everything that nature has to offer has changed. It offers little because we have weather disasters." He reviews the past year for me: "There was no spring [fruit] bloom, and the black locust was ruined. Freezing temperatures damaged the vegetation right at the start." While spring frost damage was obvious to most and the late summer heat waves are widely broadcasted on local media with health warnings, it was the beekeepers, Šefik suggests, who noted signs of drought at the very start of the summer.

"People say 'there are flowers,' but flowers cannot yield nectar! The plant withdraws all the goodness to its roots, just to survive. So, yes there are flowers, but they don't flow. Linden blossomed but did not flow; black locust did not bloom at all. And now, another ten days of drought are in the forecast."

A drought has been ongoing since July all across the country, its effects exacerbated by extreme heat waves that set in at the turn of August. Then, just days before our visit, this far-western region of Bosnia was hit by a fierce windstorm. The rain descended in punishing showers, rapidly, like a raid, and withdrew just as quickly. When the skies cleared, the air temperature soared, and the moisture the soil and plants had received quickly evaporated.

"Here it's not too bad," Šefik says of this particular apiary, "because there is all this [morning] moisture, but it isn't enough; the plants can't secrete. By now, we should have here [flowing] thyme, wild oregano, wild mint, some seventy species of blooming plants, as well as honeydew on oak leaves. With that sort of food, she'd live well, but you see how it goes. She cannot bring in the essentials let alone build up winter stocks."

Šefik is bottle-feeding the last hive in the bee yard. The syrup rushing through the bottle neck into the feeder is crimson colored by rosehips and an herbal blend of tea, which the beekeepers added to the sugar to enrich the poor food offering to the bees with vitamin C and other herbal compounds. In response to the rush of sounds and scents that the feeding introduces, the hive hums restlessly.

It takes the bees roughly twenty minutes to locate the source of nectar in the hive, Šefik says. After that, they calm down. Foragers set to work at the feeder, imbibing the syrup and processing it hundreds of times between their proboscises and honey stomachs, one drop at a time, before the sugar-based honey, enriched with bees' enzymes, is stored in the combs. In the meantime, other foragers will search the area for pollen to collect, concoct it into pollen beads, and store it in cells for further fermentation. Šefik and Senada pack the empty bottles into sacks to reuse the next morning for feeding the bees at their second apiary, situated closer to the region's chestnut forests that attract traveling beekeepers from across the country to this westernmost part of Bosnia. Chestnut, too, failed to flow this year.

"Beekeeping is now becoming a science," Šefik says, getting ready to cut the grass at the apiary. Senada, in the meantime, is preparing the bees' drinking fountains. Large jars are filled with water, their tops covered with a clean cloth that is tightly stretched and secured by rubber bands, then turned upside down so the air pressure seals the water in place. A dozen jars hang from the low branches of a tree, reflecting the garden and sky, like a mobile sculpture. The bees suckle the water that drenches the cloth at the jars' mouths.

"A human who does not know how to adjust to nature can no longer keep bees," Šefik adds. "There could be another year like 2015, when God has gifted us with an amazing flow, and everyone with bees, whether they know about apiculture or not, managed to harvest honey. But it's questionable whether such a year can repeat, the way things are going. Bees have to be on the wheels, but not everyone can head up to the mountains! Besides, bees should be distributed everywhere for the sake of the country's biodiversity."

The science of beekeeping that Šefik describes and shows through practice is a science of experience. Its method amounts to a special attentiveness to the bonds between insects and their plants, and close contact with the species in their care. Apiarists' observations glean insights into subtle, present changes such as early signs of drought that others can miss. They see hidden precarity where others see plenty, deep trouble brewing in the starved roots low beneath healthy-looking blooms.

In the course of the seasons and over the years, such close observations discern trends and afford foresight about worrisome futures. Ten years ago, Šefik foretold the scarcity of honey that has now become the norm. His prediction for the coming decade is at once more total and more vague—"everything will change." Instead of foreseeing the details of climate change effects in the near future, he considers the signs of disturbance currently mounting at hand and their implications: consistently bad weather, its stress on the plants, and the compromised health, strength, and longevity of the honeybees.

The beekeeper's science of experience is an applied science, bent on solving practical problems and guided by clear short- and long-term goals, though such goals range from keeping the bees alive through dearth to giving in order to be happy.

The seasonal nature of apian and plant lives requires that beekeepers plan ahead. The summer food determines the vitality of the next year's generation of bees, plant seeds' ripening through the fall foreshadows forage for the coming spring, and the health of trees' wintering roots safeguards the prospects of future nectar flow. Adjusting to what is becoming of nature entails maneuvering present predicaments with apicultural and typically forward-looking responses.

The immediate response to an emergency, such as feeding the bees with sugar syrup, shapes forthcoming apian bodies and collectives. For this reason, the emergency food is not just given in a timely manner (to spare the winter bees hard work), it is also enriched by teas and fruit the apiarists have cultivated, collected, and cooked. Planting for the bees' sake to secure pollen and court future nectar chances is a longer-term, ongoing venture that the beekeepers undertake and eagerly promote.

Fig. 3.1 Through the drought

## *Planting for Birds, Girlfriends, and Other Insects*

"Look, anyone who has any idea of what's going on in nature should plant at least five hundred woody plants."

"Five hundred?" I ask.

"Aha." Šefik nods, smiling, pleased at my reaction. He likes to impress others with the scale of his vision. Considering the range of the couple's planting, however, the large number does not seem exaggerated. "Senada and I have planted some fifteen hundred trees, across our lands but also in the surroundings, wherever they'll be of use."

The apiary around us is shaded by many fruit trees, and the meadow fenced off below the hives blooms in all hues. The native plants and bushes, common buckthorn and blackthorn, guelder rose and wild rose among them, mix with the new arrivals to the region, like goldenrod and tall, densely blooming mullein. The beekeepers had eagerly planted these in their apiaries and had cast their seeds through the neighboring landscapes. Mullein, in particular, they collect to prepare medicinal tea blends at the herbal pharmacy they run out of their home on the outskirts of Cazin, a nearby town.

Their pharmacy sells honeys and other hive products; a rich assortment of herbal teas, tinctures, and ointments; and their own line of nutritional supplements, including aloe, beet, and chokeberry juices. Their professional interest in herbs, plants, and power foods grew alongside their apicultural practice, and the grounds of all three of their apiaries host rich herbal gardens, berry patches, orchards, and plots with staple vegetables. Ornamental plants are everywhere, mingling among tomato and pepper stalks and intruding on plots of echinacea and marigolds.

The small bee yard by their house in town is tucked into an amateur botanical garden teeming with hundreds of species of plants that the two have painstakingly curated over the years. Some, like potted aloe vera plants, are essential ingredients to herbal remedies for their pharmacy's stock, but there are many more species thriving there that the beekeepers simply enjoy growing. Plants that are attractive to the bees are always the favorites, especially the blooming exotics like passion fruit vine and banana palms. Alongside them are weeds like common comfrey and purple deadnettle, which come in on their own, Šefik says, and are left alone. "Because, you know who comes to you?" he once asked me, then quipped before I answered: "The one who loves you." The plants' flourishing, I hear Šefik saying, shows off their attraction.

At the beekeepers' two apiaries in the countryside, trees are the most prominent plantings. The apiary that hosts us this morning is lined along the edges by dense, ruffled hazelnut shrubs. Their catkins offer up the first rich pollen loads to the bees as early as February. Leaning against them, tall and tangled, are raspberry and blackberry brambles, whose blooms well up with nectar in June. Apple, cherry, and pear trees of native and hybrid varieties blossom throughout the spring before the black locust forests in the area light up with their droopy blooms. Native and red walnuts, whose seedlings Šefik brought from Turkey, bloom with rich pollen tassels in April.

"I've planted fifty linden trees here, and now my bees can visit them. I plant *brekinja* and *Koelreuteria* at each apiary. Even when chestnut forest flows, my bees are found foraging on *Koelreuteria*; that's how much they love it." Brekinja, known in English as "wild service tree," is a native cherished for its medicinal fruits, bark, and, among the beekeepers, its nectar flow. *Koelreuteria*, originally an eastern Asian ornamental species and long at home in cultivated landscapes across the world, is new to Bosnia and rare. Flaming with extravagant orange blossoms, it is a notable bloomer among many other imported melliferous (honey-yielding) ornamentals on the couple's premises: pink euodia, blooming in a fireworks pattern; catalpa, which yields white puckered florets with purple freckles; Japanese pagoda tree; and tall Paulownias with trumpet-like flowers. Local beekeepers are tree collectors. Through collecting and caring for the foreign species of trees, they have developed a new lore, discovered new preferred species, and determined tolerances to local conditions; they can now recommend and exchange saplings with growing expertise.

Šefik and Senada are especially enthusiastic about planting. Beyond the land they own, they also plant trees across expanses of private land that has gone wild and whose owners have long immigrated to Europe, the United States, and Australia. The couple introduces tree seedlings within foraging distance of their bees and whenever they notice signs of soil erosion and suspect risks of landslides. They cast widely wild seeds of bloomers attractive to honeybees.

They also propagate saplings of the nectar-yielding trees and offer them to neighbors, fellow beekeepers, and the many people who seek the couple out for herbal remedies, hive products, or advice on beekeeping. Whenever we visit, they pack our car full of potted gifts—flowers and tree seedlings to plant at our apiary—and sachets full of seeds.

At their apiaries and beyond, this couple is matchmaking between honeybees and their partner plants. Senada points out the bees gathering pollen on great mullein this morning, and Šefik is glad: "Whenever I see them bringing in pollen, I'm happy and relieved." Nectar aside, he says, they pray to God there is enough pollen for the bees to collect. And they deliberately plant to foster pollen prospects.

Pollen in the hives is an extraordinary artifact. Deposited into special cells and mixed up with honey, wax, enzymes, and microbes from the bee's mouth and foregut, pollen ferments to become an elaborate food supply rich in vitamins, lipids, free acids, and essential nutrients. In English, it is called bee bread (*perga* in Bosnian-Serbian-Croatian [BSC]), though it is fermented rather than baked and is packed with proteins rather than carbohydrates. Nurse bees eat it along with honey to produce bee milk with which to sustain the hive mother (the queen) and the brood.[5] They also redistribute the pollen nutrients to other hive mates, complementing their honey diet. Bee bread increases bees' longevity and supports the hive insects' lifelong physiological changes, through which bees graduate from different social responsibilities, starting with nest cleaning at the youngest age to foraging in adulthood. In addition, while propolis—tree resins that foragers collect and work into "bee glue"—supports hives' social immunity, pollen has been found to boost individual insects' resistance to infection by pathogens and to promote honeybees' innate production of antimicrobial peptides.[6]

"Go on, girlfriends," Šefik says to the honeybees landing with pollen loads on their hind legs at a hive's entrance. Then he pauses to explain himself: "They aren't just bees to me; they are my girlfriends."

Bees aside, Šefik points out that their planting ventures serve other insects and birds. Mulberry trees, in particular, he plants for the wildlife, as the trees generously yield the fruit that is relatively unappreciated in the local food culture. Birds feed on the berry patches as well. Blueberries quickly withered in the high heat of 2022, but the harvest was enough to share: "I love it when birds eat here, and we should be leaving enough [of crops] for the birds and ants and everyone else. People think that all this is ours alone. Well, it's not! Properly ours is only as much as we can eat and drink, not more than that."

Planting is one way of relating to the world of many species, the world humans serve rather than own. Šefik speaks of this using keywords like *giving*, *love*, and *happiness*. We ought to give something every day, to be happy, he said. Not the usual give and take, this giving expects no returns of equivalents, but other species' thriving is the sign that love works in the world: it achieves an attraction between the species and yields pleasure.

With the bees, his girlfriends, the relationship is a serious commitment. Šefik says he loses sleep if he feels that the bees are lacking something he could provide them with, such as food, water, or herbal medicine. But the efforts at planting for the bees' sake are increasingly foiled by the strange new weather. The familiar plants' seasonal lives can no longer be anticipated. None of the trees, the countless plants in the gardens, or the bloomers they sowed in the wilds manage to secrete enough nectar to sustain the bees.

The near future, Šefik thinks, will only be more troubling. He foresees severe droughts. "The time is coming," Šefik says, "when every single plant will have to be irrigated."

## *Reading God's Signs*

The next morning, we repeat the feeding routine at the beekeepers' second apiary. The environment here is much drier and the heat's toll more visible on the plants. With nearby streams dried up and the grounds lacking morning dew, the bees are entirely dependent on the water fountains the couple refills every second day.

Šefik is thinking aloud: "We are now headed to the point from which we shall not know our way out," he says, "all because of our modern ways of living." Šefik's concerns are grounded in the apiaries but extend well beyond them as he worries about what life will become with the loss of biodiversity and climate change. He blames modern agriculture, the fossil-fuel industry, pollution, and consumer culture, the anthropogenic pressures that are typically implicated in the global environmental crisis.

What Šefik envisions, however, is a point of no return, the eventuality of a postsustainable world that few contemporary environmentalists are willing to contemplate. Environmental philosopher John Foster calls this a "grievous" future whose shape we cannot know well in advance, and neither modern technology nor our present imagination can help us live toward or through it. The way forward, Foster suggests, begins by staring frankly at the abyss that opens at the thought of an imminent catastrophe, the possibility of our species' extinction, and the eventuality of our personal death. Foster argues that confronting finitude and environmental tragedy can help cultivate a new species of hope that taps into the deep "dark" part of the human self, which we intuit even as it eludes our grasp. When we retrieve the inner human wildness whereby embodied beings think and live as an organic part of the ecologies they inhabit and that ground them, hope arises from that, unwilfully, and we can strive to "keep the human torch alight."[7]

Šefik's understanding of global environmental disaster draws on his intermittent participation with internationally funded projects in local biodiversity conservation and sustainable development as well as from his personal observations through apicultural and herbalist practice. At the same time, his idea of what nature is and means to the human self draws on the Revelation, which the Qur'an describes as the light that illuminates one's course through life: *We have made it a light by which We guide whom We will of Our subjects.*[8] God's name is the Light, *al-nūr*, as in the famous Qur'anic verse that begins with *God is the Light of the heavens and the earth.*[9] Revelation, in other words, is meant to illuminate human self-understanding, those corners of inner being that would otherwise stay dark precisely because they would remain within the shadow cast by humans' lone, self-wrought self-perceptions.

As Šefik put it earlier, nature is the realm "of goodness, which is the God's gift." The present trouble stems from humans breaking connections with the gift and its Giver.

"We have distanced ourselves from the nature," the apiarist says, "and the human has to be connected with the cosmos, with one's Lord, every second, no matter what you're doing, because the inputs are coming from above, if only you know how to receive them. It's like with bees; they're receiving their revelation."

Nature is a fraught concept in contemporary Euro-American eco-thinking. Some deem it lost under the sway of the pernicious influence of late modern humans while others suggest that nature has always been mixed up with human influence, although the Western mind and modern science presumed it to be an objective realm wholly separate from culture.[10] Others are recovering the idea of the wild as an alternative to nature to capture unforeseen tendencies in nature or culture that are ecologically or politically promising.[11] In any case, conversations that revisit or problematize the concept and domain of nature evolve around relationships between human and nonhuman entities, be they biological species or any species of things, from atmospheric elements to cosmological beings, which conceptually adventurous scholars are now also considering.

Šefik's notion of nature, however, is colored by the Qur'anic understanding of the cosmos as everything other than God but nothing apart from God, who brings it into existence and sustains it from one moment to the next. Everything there is, is a subject of God's mercy and a pious Muslim subject in the sense of being gladly, adoringly surrendered into divine service, as per the following verses, for instance: *Then He turned Himself to the heaven, when it was smoke, and said to it and to the earth, "Come forth, willingly or by force!" They said, "We come in willing obedience."*[12]

For humans, the natural world, the cosmos at hand, is the place and the means of relating with other divine subjects. Giving service and showing care to those who themselves are doing service to God and humans, for the love of God, is the intermediary way of relating to God through good works. And the practice of doing good, the Qur'an repeatedly says, goes hand in hand with having faith, as opposed to simply professing it.[13]

God's nurturing relationship with the world is still more intimate insofar as the divine attributes, known in the tradition as God's Most Beautiful Names, are manifesting in the sensuous world. God reveals Himself through properties and events at all times so that everything in the material world encounters humans with the latest news of God. These signs are meant as occasions for reflecting on and for getting to know God, who never ceases to speak. A Qur'anic verse well loved by Muslims says as much: *And if all the trees on earth were pens and the* sea ink, with seven seas to replenish it, the words of God would never be exhausted.[14]

The connection with God via nature is reflective but also more existentially vital to being a human than is reflection understood in the usual sense of the word as a discrete, mental activity of a thinking subject. The human, in the Qur'anic sense of the word *al-insān*, is the being with and in the name of God. While all things are divine subjects, and their glad service to God makes them Muslim, being a Muslim is a prerequisite for humans becoming an *insān*, a being drawn toward God. And while everything is in a constant process of change, and humans are particularly prone to differentiation, personal constitutions being diverse and each person constantly subject to passing moods, the divine self-revelation offered to humans through the pages of the Book or the beings in the world is incessantly singular. *Insān* is a being underway.

Revelation sustains the honeybees as the foraging, pollinating, honey-making, prophetic species whose relationship with God is faithful and always accomplished. For humans, Revelation is meant to guide the process of what the local Sufi thinker Mustafa Čolić calls *insanizacija* ("insān-ization"), which I take to mean getting to know one's self as not god as the precondition for getting to know God. A prior receptiveness to divine self-revelation, however, sets the stage for self-knowledge, since a self reflects nothing but God Himself, though the reflected image depends on the surface of the mirror. The self-image veils God, and veils can be more or less luminous or revealing.

For Shaykh Čolić, the process of becoming an *insān* also hinges on the intention and desire to walk in the footsteps of the Prophet, whom hadiths quote as saying, "I have been sent to perfect good character." Good character (*akhlāq*) shapes the whole range of polite conduct, from handling water to responding kindly to insults to meeting God. Precisely because humans, unlike bees, can lose and must struggle to maintain their connection with the divine, such loss is disastrous not just for the individual but, Šefik implies, for the world.

Signs of coming disaster are legible on the pages of the Qur'an, God's Book, and discernible across the open book of nature. Šefik refers to the signs with the loan word from Arabic, *išaret* (*ishāra*): "Before the End [*pred Kijamet*] we shall witness all the signs, and they'll be greater each day. We cannot know when, God knows [when is the Hour], but God's Book speaks of it, and the portents are here." Šefik stands still, taking time to think. He struggles to find the words. "It's up to God's will, but each day we'll ... I won't say that we'll be suffering [more], but, for sure, we will be realizing further what we have done [to bring this]. Each day we'll be grasping deeper the meaning of what's written in the God's book, to the extent that we know how to interpret it. The great signs are already here ... we speak of it as 'climate,' and the trouble *is* about climate, and we have caused it."

At this point, Šefik changes his tone and gives a broad, sweeping critique of the carbon economy, unsustainable industrial growth, pollution, xenobiotics that are making landscapes and bodies toxic, global politics of complacency, the tendency of the general public to ignore the "alarm, blasting to all of us." Šefik vents his grievances with sadness, not anger. The cheerful man, not prone to complaining, however, soon shifts his mood and says: "If we want to live better on this planet, we each should plant at least five hundred woody plants, to help out. Because it's so nice here, right? Being here by the forest. This is a God's gift."

Both the Revelation and Šefik's firsthand observations yield signs of the coming catastrophe: nectar dearth, severe droughts, the end of life as we know it, and, imminently, the end of the world. Wrecking the planet is the modern human industry in the widest sense of the word as well as the human distance from nature and God, whose gifts manifest in nature. And yet Šefik promotes tree planting.

"Why should we plant?" I ask.

"Plant a tree even if you knew you'd die in an hour," he says, repeating the prophetic advice with a notable difference: the eve of the apocalypse, in his telling, is the eve of one's own death. The conflation of death and apocalypse is not incidental, as the primary Islamic sources and their commentaries deliberately thread together the finitude of the self and the cosmos.

A verse in the Qur'an cites God: *What else can they* [the people] *expect but that the Hour should come to them suddenly, and its portents are already here. And what if it comes and they are unaware and have not prepared for it?* Mustafa Čolić glosses the Hour in this verse as both the private and the general apocalypse (*smak*), the perishing of both the macrocosmos and microcosmos.[15]

Fig. 3.2 Off to the next task

Both versions of the hadith circulate in casual conversations, and the point they jointly make is that neither the end of one's life nor the end of the world intimates futility. Planting that tree is not meant to stall one's death any more than it is meant to save the world. Planting is about doing the good that keeps the world whole, about being through giving, which serves the world of God's creatures and earns God's pleasure.

"Now, let's go harvest chokeberries," Šefik says, picking up a bucket. "They are ripe." Senada is already way ahead of us, gathering mullein flowers in the meadow. Šefik heads to the chokeberry bushes, and my sister and I follow.

## *Something Is Off*

"It needs another ten kilograms of food," Mehmed says, closing off the hive he had been showing us. The hive holds the swarm he had caught at his apiary earlier in the summer. He is pleased with how the bees have developed thus far, except that the honey stocks are not yet ready for the winter. It is early August, and we are in central Bosnia, some three hundred kilometers east of Šefik and Senada's apiary. Mehmed is holding off on artificial feeding for as long as there is a chance that the meadows might flow. Taking off the yellow beekeeper's shirt and the veiled hat he wore while handling the hive, he explains: "These hills are full of different nectar-yielding plants. There is plenty of forage, and it's blooming. There are north-facing slopes and south-facing slopes. In short, whatever exposition the plants like, they can find it here. If only other factors come together, the flow shouldn't fail. But, by God, in the last years, something is off." He shakes his head and looks off into the distance: "Something is off."

Mehmed's apiary is in his home's backyard, which extends across several acres of land, well planted with fruit trees and berry patches. His appraisal of the surrounding hills conveyed a beekeeper's and an orchardist's consideration of the microclimatic conditions that are fostered by the angles of the slopes and their orientation, the plants' exposure to the sun and winds, daily temperature oscillations within seasons, and the soils' capacities for water retention. By all counts, it is a good place to keep bees. He inherited the orchard from his father, a traditional beekeeper, and has introduced many new fruit and flowering trees over the years.

Some twenty hives are in the bee yard, and the rest have traveled the country in pursuit of honey since the spring. They are currently at two different locations in the highlands, poised for the summer heather, the region's last major nectar source.

Mehmed's traveling apiaries fared poorly. Should the summer heather fail, he will bring them home early. "At least here, I didn't have the situation where I had to urgently intervene to rescue the bees on the brink of starvation," he explains. Midsummer, he had to evacuate his bees from the spot in the mountains to which they traveled, as in any other year, to take advantage of the later blooming season and the diverse meadow forage that flourishes at higher altitudes. Finding the hives dangerously empty of food on his field visit and seeing no near signs of nectar flow, he rushed the bees back to his home base, fed them homemade pollen patties, and then took the hives to the mountains southward.

On average, a honeybee society consumes 150 kilograms of food, nectar and pollen combined, a year. The fact that, lately, they can hardly manage to collect that much is alarming, Mehmed says. "Someone may think the lack of honey is the beekeepers' problem. But shouldn't we all be worried?" he asks me, laughing nervously. "In these times, the bees cannot sustain themselves without human assistance."

The assistance Mehmed speaks of is the apicultural practice itself, which, in his words, has become "a struggle to keep the bees alive." The practice involves careful management of the hives to ensure the bees' health and strength, but the main forms of aid are itinerant pursuits of nectar and cultivation of bee-friendly trees and plants.[16] When all else fails, there remains artificial feeding. Although Mehmed supplies the hives with the best possible food alternatives—sugar syrup enriched with herbal teas and patties he prepares from scratch—artificial food is considered a rescue response and a sure sign of an apiculture in crisis.

Mehmed has deliberately planted the apiary's surroundings to ensure that sources of nectar and pollen are available across the growing seasons. When I ask him what he has planted for the bees, he simply replies: "Everything! For each season she [the bee] has something to pick." He gestures in different directions across the area and lists the bees' notable partner species. Among them, thickets of hazelnut ("you can't breathe here in the spring from the amount of pollen in the air") and groves of Cornelian cherry, an early and opulent bloomer, with nearly a hundred of its trees growing dense, low, interlaced canopies within walking distance from the apiary.

In the orchard itself, he has grown native and hybrid varieties of apples and pears, deliberately selected for the different timing of their blooms. Theoretically, the fruit blossoms should last at least twenty days on these premises. "Because here, there are the earliest, the early, the midseason, and the late bloomers. However, nowadays, within a single week." He claps his palms saying this: "It's all gone! Or else, the latest bloomers bloom first." Mehmed takes a thoughtful pause, then adds, "Inexplicable!" and wipes his silver-rimmed glasses. He is a composed, cheerful character. In his early fifties, and looking younger than his age, he has a way of making the gravest statements with a small nervous laugh that intimates at once his unease and bafflement.

The order of flowering is upturned, Mehmed notes, and not just with fruits. He mentions as an example brekinja, wild service tree, which he is fond of cultivating on his land and in the nearby forests.

Fig. 3.3 Black locust, nearly exhausted

"We know the sequence in which plants here develop, what comes before and what follows next, but that order no longer holds. Now brekinja blossoms before black locust!" The service tree is an important bloomer, and its flowering was timed auspiciously to follow on the heels of black locust and so fill in the gap until linden, the next significant arboreal flow.

The familiar blooming sequence of other local trees, such as the maple, as well as the different meadow flowers is likewise disordered. The beekeeper is relearning the disrupted rules of plant phenologies. The native plant species and the foreign ornamentals that have been at home at Mehmed's apiary for nearly two decades are throwing up surprises. He sees the ornamental euodia bloom as generously as ever, but there seems to be no flow.

"We know from literature and from years of experience, that there are ideal meteorological conditions for different plants' secretion. But this year, too, it happened: the conditions seem ideal, but there is no honey. . . . We have strong, healthy bees, capable of foraging, but there is no nectar. The problem we're seeing in the hive is only a consequence of something that is going on in nature. Something is disturbed there."

He takes my sister and me for a stroll through the meadow far behind the hives. Grasses are shot through with wild plants, white and purple wild clover, spiny restharrow blooming pink and armed with thorns, lacy yarrow, celestially blue chicory, and many more.

"Take today, for instance. Today we have ideal conditions for nectar flow. Did it rain? It rained, so there's moisture. The temperature is right, the flowers look good—they are perky, as you can see, not dried up—but if we open up a hive, we'll see there's no inflow," Mehmed says, his frustration building as he speaks. He pauses, laughs, then falls silent.

"I don't know in which direction things are going," he says finally, "but we, the beekeepers, aren't going to give up. This isn't just about the work; it's the way we live. And I don't want to and cannot live without her."

## Halal: Giving while Forgiving

Honeybees run deep in Mehmed's family. He bonded with them as a child, by his father's side, in this very orchard, which, back in the 1960s, hosted bees in traditional, whitewashed wicker skeps. The boy watched the summer skies explode with swarms. He rediscovered them as a young man enrolled in Islamic studies at the university while studying to become an imam. *And Your Lord revealed to the honeybee*—the Qur'anic verse begins. Reflecting on this verse and the famous fourteenth-century commentary by Ibn Kathīr, it dawned on Mehmed that foraging honeybees show God at work at the apiaries, guiding the bees while making Himself known to humans. The verse on the bees ends with *In that is the sign for people who reflect.*

On his days off from the battle trenches in the 1990s, Mehmed spent time with his wife, their young daughter, and the hives in their backyard. Intense, rushed days, he remembers, when the bees matched him up with the seasons and their honey harvests to the sweet life that the war besieged. On all other days through wartime, his wife, Fahreta, picked up the tasks around the apiary and their older daughter joined in as soon as she could stand on her own, the adult-sized veiled hat swaying on her toddler head like an overgrown mushroom cap on a slender stalk. When, after the war, Mehmed resigned from his post as a local imam, honeybees became his full-time vocation. Years later, the apiary remains a family affair.

Ever since their younger daughter, Dalila, a teenager at the time, was diagnosed with lymphoma, the hive products—honey, pollen, propolis, and royal jelly—became staples of the family's diet,[17] first to nourish the girl's body through chemotherapy and promote convalescence, then as everyone's prophylactic. To grow power foods for Dalila and secure produce free of contaminants and agricultural chemicals, the family's gardens were expanded.

They added patches of chokeberries, raspberries, and strawberries, pots of aloes and live-forevers, beans, and squashes. The family still largely eats what they grow. Roses, however, grow everywhere throughout the gardens and the orchard because, as Fahreta puts it, "the human finds room for pleasure, everywhere." The expanded food garden, too, is the honeybees' heaven, and pleasure travels with them to its every corner. As a young adult and master's student, Dalila still helped around the apiary while carrying her close interest in bees into the studies of microbiology; her thesis project tested the antibacterial properties of bee venom.[18] The beekeepers' older daughter, employed as a pharmacist, has no time to spare at the apiary, but her daughter, Najma, practically grew up with her grandparents' honeybees.

Najma once showed me her littlest finger, excitedly explaining something in the tongue of a toddler I could not understand. Her grandmother translated for me: "She is showing you where a bee has 'kissed' her." The adults turned the child's first sting into a lesson on the human-apian relationship. Across the species, love is love, though it gives and feels differently.

Local beekeepers, and especially those who, like Mehmed and Šefik, are planting for the bees' sake, are engaged in close, loving relationships across the species' differences. With bees, we do not see eye to eye. We cannot exchange gazes between our respective organs of vision—the cyclopean three eyes on the bees' forehead and the compound eyes on either side of the bee's face, with thousands of lenses covered with stalks of hair, standing tall, at the microscopic scale, like prairie grasses, grasp a world of light, movement, and shape alien to us. Our close contact presumes avoidance of touching. Our fingers are too coarse for their frail, exoskeletal frames, while our skin hurts at their stings and clutches their barbed stingers irretrievably, rupturing the bees' abdomens. The local beekeepers' love for the bees does not count on these human tokens of intimacy—gazes and caresses—any more than it expects an apian recognition of the beekeepers' presence, service, or affection. The beekeepers' love does not expect a reciprocity, an exchange of likes, a tit for tat, kisses for kisses, a care in return for care. Instead, the emphasis is on human giving.

"The bottom line is this: What's up to me, I'll do it. Whatever I can do for my hives, I'll do it. Will they give me back anything? They don't have to. It's my way of life. Perhaps next year they'll give something back, but they don't have to. *Halal im bilo* [I forgive them]. It's enough that I come here, fill up my soul by the hives, and it is plenty." Mehmed points to a canvas garden chair off to the side of the hives. He likes to sit there in the evenings, he says. As the crickets fire up the summer's usual, the foragers are making their way back to the nests, and the hives are breathing out the scents of wax, tree resins, and the nectar becoming honey.

Halal im bilo means that what Mehmed gives to bees absolves them of any debts. They owe him nothing in kind—honey or hive products—nothing in turn. A Qur'anic Arabic term, halal, refers to the category of acts and objects that are permissible to Muslims; in other words, their enjoyment is rightful. Bosnian Muslims tend to call out halal on a range of transactions, be they of money, words, or deeds, to emphasize the gladness of giving or forgiveness of any eventual slights or imbalances. Halal resolutely renounces grievances—someone who has been hurt or insulted shows kindness and virtue by forgiving. Above all, halal counts on God witnessing and keeping track of even the smallest act. Those who hope for God's mercy ought to be forgiving; those who depend on God's generosity ought to be giving. Whereas this forgiveness is most explicitly emphasized in relationships between humans, I also see it tacitly informing Bosnian Muslim ecological sensibilities.

Mehmed is well aware that the Qur'an describes everything in the cosmos as employed with giving service: from the moon to the livestock. Attentive readings of those verses emphasize the fact that humans do not benefit from this service as rightful owners.[19] The world is not at humans' disposal. Rather, humans are utterly dependent on the world's creatures and are especially accountable to God for the benefits and blessings they draw from a world that ultimately belongs to God.[20] While all things are employed by God and render service to others, including humans on God's behalf, human living, by default, accrues due to all things that make living feasible. Mehmed gives to the bees, forgivingly, while all along the bees in his care are indebting him through service.

The generous God promises eternal rewards to those who give and forgive. But because He is above the condition of owing something to His subjects at any time, God rewards the good works of all creatures instantly: with pleasure. In the words of a nineteenth-century Sufi, Said Nursi: "Know that out of His perfect munificence, Almighty God placed the reward for work within it. He included the wage for work within the work itself. It is for this reason that in their particular duties, which are called creative commands, animate beings, and even from one point of view inanimate creatures, conform to the dominical commands with complete eagerness and a sort of pleasure. Everything from bees, flies, and chickens to the sun and the moon carry out their duties with perfect pleasure."[21]

God's creative command engenders things—*His command, when He desires a thing is to say to it "Be," and it is*[22]—while divine generosity ensures that God's *desire* comes through with the command and that things come forth with pleasure.[23]

The pleasure that Mehmed speaks of is sensuous: the peaceful, leisurely moments steeped in the sounds, scents, and textures of a summer evening. The pleasure is also more than that since Mehmed uses the word *soul* (*duša* in BSC), which is a complex term in Islamic sources. At the very least, the human soul is conceived as a double. It consists of a *nafs*, a subtle reality that craves and enjoys fleshly as well as sensuous or aesthetic qualities of this world (and the next), and *rūh*, which is the divine reality, the spirit breathed at the secret heart of a human being. *Rūh* craves divine attention and thrives on the deeds that seek to earn it. The use of these terms in Islamic sources varies; in everyday Bosnian Muslim conversations, duša often signals both nafs and rūh, while the use of Arabic terms usually serves to mark notable distinctions between the two.

*Nafs*, in particular, is the part of humans that inclines toward disorder and necessitates the prescriptive command, which amounts to divine instructions on how to distinguish good from evil, what duties to perform, and what to avoid in order to earn God's pleasure. Humans (along with jinn) are the only species obliged by, as well as free to disobey, the prescriptive command.

Being generous is a highly praised virtue in Islam and a practice that counters the selfish tendency of *nafs*, which by nature is self-centered. Citing a hadith, Shaykh Čolić writes that a generous *insān* is one who complies with divine prescriptions while giving dues to God and, by default, to God's beings, whose dues are the subject of consideration in the giving of alms or taxes. The apiarists described in this chapter make it clear that generosity is giving due to honeybees and plants being entrusted in their care. The reward is immediate and fills up the soul.[24]

### *If You Saw the World Ending*

Our visit is drawing to a close. We are standing at the edge of the bee yard by Mehmed's garden chair, the seat of his evening repose. "What can we do? How much can we do?" he wonders out loud and in the next breath adds: "I do what I can. See there?" Mehmed asks, pointing out a cluster of tall trees nearby. "Those are linden trees that I've planted." Five, he counts. "There are as many up there, and then more the further you go," the beekeeper says, nodding toward the forests that skirt the surrounding hills.

The trees Mehmed points out are some twenty years old, with beautiful crowns naturally rounded as is typical for linden. I imagine they turn gloriously rich and fragrant in July, when they bloom. Along with all the other trees in Mehmed's purview, the linden has not been secreting as expected and does not reliably aid the apiculture in crisis. The trees Mehmed plants and tends to are being uprooted from the seasons, their life cycles and species tendencies disjointed from the ambient conditions. Nonetheless, the prophetic advice to plant is deeply ingrained and remembered precisely at the moment it seems that the world one is planting is beyond redemption.

"If you have a sapling in your hand and you see that the apocalypse [*Kijamet*], has begun, plant it," Mehmed says, clutching his fist as if it gripped a tree stem. "You are seeing all the signs of Kijamet? Well, don't give up, plant it! What does that sapling matter, if the world is ending? It matters. Plant it!"

He gestures toward the first hive at the apiary, the one we opened earlier today that houses this year's swarm, and says: "Will I live long enough to see these bees forage? Will they get to find honey next year? I don't know, but I'll do my best to turn them into a strong community, and next year, I'll carry on, if I can, inshallah [God willing]. That's the bottom line."

### *Planting in the Paradise*

Lacy phacelia flowers color purple the field that stretches ahead of us, then rise into a hill slope suddenly, like a hiccup that interrupts a flowing conversation. The higher the hill gets, the more scarce and withered the blossoms. "The common ragweed has taken over," Adil says with a sigh, pointing out the hardy blades of an invasive species, an infamous allergenic, which is a recent arrival to the country.

Since the spring of 2021, Azra and I have followed Adil's efforts at cultivating lacy phacelia, the plant that promises abundant nectar flow for the bees in his apiary. He had procured the seeds from abroad, plowed the field with his small tractor, spread out manure, and, finally, in May, cast the fine seeds. We had helped out with sowing. Even before the first flowers emerged, as early as mid-June, he noticed that common ragweed had overwhelmed the crop. In his words: "The weed won over. It's wild, invasive, and has a tougher root, so it drains the nutrients and the moisture."

As temperatures soared throughout June, the phacelia blooms quickly burned, except at the foot of the hill, where we meet Adil on this visit in mid-July. Morning dew refreshes the plants there, and the shade from the apiary's many trees keeps the patch relatively cool through early afternoon.

As a young man, leading a faith community in a northeastern city mosque, Imam Adil once dreamed of a fabulous garden. In that dream, which Adil vividly remembers years later and relates to me with glee, he flew across fabulously lush landscapes, as light as a bee, finding each place more astounding than the one before it. The dream gave him a foretaste of paradise for sure, he says. Since Adil retired from the post of imam, he and his wife returned to his paternal village by the name of Rajska, which translates to Edenic and fittingly describes the places' verdant, faraway charm.

Rajska is off the main road, deep within the woods, and at a safe distance from the commercial fruit farms that are infamous among the beekeepers for being highly contaminated by synthetic insecticides and herbicides. Adil has kept an apiary here throughout his adult life, visiting the bees on his days off from work. Now he lives full-time in the house by the apiary, which is an eclectic collection of aging boxes of all types and sizes. Rows of hives are screened from each other by high-climbing blackberry and raspberry vines. Shrubs of roses and jasmine and hollyhock in many hues make up the outer lines of the bee yard, which to the left grows into an orchard and up front becomes a thick vegetable garden pressing against a greenhouse. When he is not with the bees, Adil keeps his hands in the dirt. He tirelessly plants trees, though his wife and adult children worry that Adil, now in his seventies, is overextending himself. He undertakes these new planting ventures under the pretext that he plants on behalf of others—he tells me with a wink. With his wife's help, Adil cultivates food, berries, and flower gardens. While the number of hives in the bee yard is steady at a hundred—he gives out each year's swarms as gifts—the gardens and orchards keep expanding.

The land at and around the apiary is planted with hundreds of fruit trees. Adil has a story and praise for each variety: one apple is "clever," he says, because it blooms in well-paced stages, another is "neat" because it requires no pruning. He has planted dozens of varieties of cherries and sour cherries, and each spring, he takes us to see the bloom of his favorites. Among these are Japanese cherries, which he calls "jewels." He gave saplings of it to my sister and me earlier in the spring, along with other potted gifts for the land of our apiary: a black fig tree, a flowering bush that attracts bees ("a true ornament"), dozens of raspberry saplings to add to our berry patch, and a sack of hollyhock seeds.

Several years ago, we introduced Adil to the false indigo bush. From the sapling we brought him, he has propagated others, founding a small thicket on the wettest part of his land. Now he recommends the tree enthusiastically to his fellow beekeepers and passes saplings of it along with advice on how to control the spread of this vigorous grower. By the side of the village road, on the bank of the cool forest creek, he has established a plantation of Paulownia trees.

"This is the tree of the future," he said, giving me a tour of the grounds in 2019. Adil listed the properties of the species that has been hailed globally as a fast grower and a versatile, promising resource in sustainable forestry, but he kept his focus on the paulownia's reputedly high nectar yields. By the time the tree was discovered by the local beekeepers, Adil's Paulownias were already several years old.

All around Adil's land are forested hills full of black locust and linden. In 2021, as in the years before, black locust failed to flow. Adil suspects that the trees were impeded by late frost, which also disrupted the flowering of his fruit trees. Later fruit varieties blossomed through persistent cold showers that confined the bees to their nests. Paulownias flowered "somewhat," he reports, unimpressed.

Still, *alhamdulillāh*, praise to God, Adil says cheerfully. It was the phacelia and linden that infused his hives with food in the nick of time. Everything else up to that point has disappointed: nectar that trickled in from trees and plants on Adil's land and from the surroundings could not feed the bees. At the very least, Adil felt, the small inflow of nectar, along with the pollen that the foragers managed to collect, enriched the sugar and soy patties that the beekeeper fed the bees through early summer. Before the linden and phacelia flowed, it was a "catastrophe," as Adil puts it. Now the hives have food for the rest of the summer, and there will be a modest quantity of honey for the beekeeper to harvest, which, in Adil's words, is plenty to supply his family and friends. "You see, I could live without honey, but I could not live without bees," he told me the first time we met in 2015.

We are now slowly climbing the hill planted with phacelia, looking closely at the flowers along the way, speculating how the ragweed might have gotten into the crop. Adil is wearing a floppy red hat that shades his face from the sun's blazing heat. From the hilltop, we have a good view of the village in the valley and the hills rolling into the distance. Adil points out the black locust and linden forests in the area. When these trees flow with nectar, it is a paradise for bees, he says, then shrugs his shoulders: he would not know what to do with all that honey anyhow. "I now keep the honeybees mostly for the pleasure," he says, "and for the sake of *hizmet*."

*Hizmet* is a loan word from Arabic via Ottoman Turkish that denotes service to others, which is highly recommended in the Islamic tradition. *The most beloved among God's subjects is the one who is most useful to others* is the gist of a hadith that Bosnian Muslims often tell each other, encouraging solidarity within the community as well as charitable works of any sort. Adil's use of the phrase *the community* suggests a wider meaning.

"Not just to the humans?" I ask Adil to clarify.

"Of course not!" he keenly replies. "God created those animals and has entrusted them into your care. When you do everything you can, you earn a *sevap* [reward]. For every little bee you serve." He gives a deep, hearty laugh at this, then adds: "There about five million of them here, you know."

A hundred hives, five million of God's creatures, whom Adil calls girlfriends. He describes his tender relationship with them as a service, giving to the creatures of the world their due, for God's sake. The word sevap is another Qur'anic term (*thawāb*) denoting a reward, while the Revelation makes it clear that rewards pertain to this world and the next.

Mentioned with reminders of one's imminent death, the Book encourages its readers to strive and pray for both kinds of rewards: the coveted ephemerals that make life enjoyable and the everlasting goods. Consider the following verses: *No soul can ever die save by the leave of God, at an appointed time. Whosoever desires the reward of this world, We will give him of it; and whoever desires a reward of the other world, We will give him of that, and we will recompense the grateful.*[25]

The three beekeepers I describe in this chapter tie together a common theme. Mehmed gives to the bees all he can, renouncing expectations of returns. The service "fills up" his "soul." When Šefik recommends that we humans "give something every day to be happy" (*da bi bio sretan*), he has in mind doing what makes one feel good and doing the good that rewards souls with eternal bliss. The bliss, however, is advanced in the present and is what nourishes the soul.

In the Qur'anic sense, the felicitous—translated in BSC as *sretni*—are distinguished on the very day that follows the world's end, the day of gathering: *And that is the Day that all shall witness, and We delay it only for a limited term. On the Day it comes, no soul shall speak save by His*

Fig. 3.4 The sun rises from the west

*leave. Among them shall be the wretched and the felicitous.*²⁶ The wretched and the felicitous are mentioned together throughout the Book, and it is the contrast of their deeds and ends that deepens the meaning of misery and joy in both the earthly and the eternal life.

Šefik's contemplations are kindred. To emphasize the gladness that comes from giving, he invokes its opposite with his typical boldness: "The man who is not useful to self and others is miserable."

Gladness stands for the Garden. The Qur'an quotes God: *And as for those who are felicitous, they shall be in the Garden, abiding there for as long as the heavens and the earth endure, save as your Lord wills. Indeed, your Lord accomplishes what He desires.*²⁷

The garden, however, is sowed here, in the loam of the earth. A popular hadith among Bosnian Muslim beekeepers says that a Muslim who plants a tree or sows grain reaps rewards for as long as its fruit is eaten by a person, bird, or animal. Adil quotes a hadith that is even more straightforward: whoever plants a tree, God plants a tree for him in paradise.

There is a logic to planting here and hereafter. If you plant a tree here and now in God's garden, you join hands in the divine care for the world. Even if the earth is doomed and plant nectaries are exhausted, the merit of planting rests in the seedling itself, and it will bloom in all the worlds.

Notes

1. 70:6–10.
2. Ibid.
3. In cultural traditions across the world, honey is appreciated for its nutritional and medicinal properties, some of which are being reevaluated in clinical practices and researched in international interdisciplinary networks. Peter Molan, a biochemist, has devoted his career to exploring and promoting honey's medicinal uses, starting with his article published in the 1990s that argues for reconsidering honey's therapeutic potential. Largely thanks to Molan's research efforts, medical grades of honey became part of clinical wound care. A number of influential works on the medicinal uses of honey and hive products circulated in translation throughout the socialist East during the 1980s. In Bosnia and Herzegovina, renowned journalist—and beekeeper—Nijaz Abadžić was the most responsible for popularizing international research on the nutritional and therapeutic properties of honey. His 1967 *The Secrets of Bee's Honey* remains a classic reference for local beekeepers. Nijaz Abadžić, *Tajne pčelinjeg meda* (Sarajevo: NIP "Zadrugar," 1967). Conversely, according to *The ABC & XYZ of Bee Culture*, a lead sourcebook of North American apiculture, "Honey is a sweet, viscous material produced by bees from the nectar of flowers, composed largely of a mixture of the two sugars glucose and fructose dissolved in about 17% water. It also contains very small amounts of sucrose, minerals, vitamins, proteins and enzymes." As *The ABC & XYZ of Bee Culture* points out, the major components of honey are sugar and water (the monosaccharides fructose and glucose make up some 70 percent; disaccharides, such as sucrose, amount roughly to an additional 10 percent; and the water content is anywhere from 17 to 20 percent). However, what chiefly contributes to honey's character—its flavor and aroma—are minor constituents that are extremely complex and perishable, derived from both the plants from which honey is sourced and the chemical processes it undergoes in and between insect bodies and within the comb. Studies have identified 181 substances in honey, some of which have not been noticed anywhere else in the natural world. Chemical identification of components responsible for honey's flavor and aroma has proven difficult. Many compounds are at play, including acids, amino acids, enzymes, and a range of molecules with an aromatic ring collectively known as phenols. Honey's therapeutic properties are also associated with these subtle compounds that make up so little of honey's measurable, identifiable body volume. Among them are flavonoids, phytochemicals described as strong natural antioxidants with a range of beneficial effects: antiallergic, antibacterial, and anti-inflammatory. The phenolic content of honey varies according to the floral nectar sources and is further determined by the many environmental and atmospheric factors that distinguish honey across geographic areas and seasons. Furthermore, various forms of sugars found in honey have been associated with honey's prebiotic activities. Fresh honey also contains probiotics. These microorganisms survive in honey several months after the harvest; consumed, they inhibit the growth of pathogenic microflora in the human and animal gut. Proteins, enzymes, and essential and nonessential amino acids make up a mere 0.5 percent of honey's volume and, therefore, are not significant with respect to the required human daily protein intake. That being said, enzymes are vital participants in honey's antimicrobial and antioxidant activities: they promote the absorption of calcium, starch, and sugars and serve a biological function in disease prevention. The amount of vitamins and minerals in honey is also small although, depending on the source, honey can supply chromium, manganese, selenium, and a number of other micronutrients that are important to human health but for which no daily intake values have been proposed. Honey is also a significant

source of choline, a nutrient essential for many metabolic processes that support cell repair and cardiovascular and brain functions. Honey is furthermore a good source of acetylcholine, an organic chemical with neurotransmitting functions. I hope this lengthy note deepens the meaning and value of honey, the substance that is too unique and irreplaceable to be described as simply a sweetener or a carbohydrate. For further readings, see Amos Ives Root, *The ABC & XYZ of Bee Culture* (Medina, OH: A. I. Root, 2007), 888; Peter Molan, "Why Honey Is Effective as a Medicine, 1. Its Use in Modern Medicine," *Bee World* 80, no. 2 (1999): 80, https://hdl.handle.net/10289/2059; Eva Crane, *A Book of Honey* (Oxford: Oxford University Press, 1980), 39; Lai Moon Dor Ginnie Ornella and Fawzi Mahomoodally, "Traditional and Modern Uses of Honey: An Updated Review," in *Honey: Geographical Origins, Bioactive Properties*, ed. Ruben Ramirez (New York: Nova Science, 2016), 81–98; Elena Saltykova et al., "The Effect of High-Molecular Weight Chitosans on the Antioxidant and Immune Systems of the Honeybee," *Applied Biochemistry and Microbiology* 52, no. 5 (September 2016): 556, https://doi.org/10.1134/S0003683816050136; Tahereh Eteraf-Oskouei and Moslem Najafi, "Traditional and Modern Uses of Honey in Human Diseases: A Review," *Iranian Journal of Basic Medical Science* 16, no. 6 (June 2013): 734, https://doi.org/10.22038/IJBMS.2013.988; Visweswara Rao Pasupuleti et al., "Honey, Propolis, and Royal Jelly: A Comprehensive Review of Their Biological Actions and Health Benefits," *Oxidative Medicine and Cellular Longevity* 2017 (2017): 1259510, https://doi.org/10.1155/2017/1259510; Stefan Bogdanov et al., "Honey for Nutrition and Health: A Review," *Journal of the American College of Nutrition* 27, no. 6 (December 2008): 679, https://doi.org/10.1080/07315724.2008.10719745; Abadžić, *Tajne pčelinjeg meda*; Naum Petrović Jojriš, *Pčele i medicina*, trans. Aleksandar Đerdanović (Banja Luka: Romanov, 1974).

4. The life span of modern honeybees has been considerably shortened. Among the factors implicated are contamination by chemicals, parasite infections, inadequate food resources, climate change, and stress related to intense work schedules. For research that correlates the honeybee life span with the size of the hive and, hence, the intensity of insects' collective work and development, see Olav Rueppell, Osman Kaftanouglu, and Robert E. Page Jr., "Honey Bee (Apis mellifera) Workers Live Longer in Small Than in Large Colonies," *Experimental Gerontology* 44, nos. 6–7 (April 2009): 447–452, https://doi.org/10.1016/j.exger.2009.04.003. Life expectancy also determines foraging behavior: bees with shortened life expectancies begin foraging earlier. See Michal Woyciechowski and Dawid Moroń, "Life Expectancy and Onset of Foraging in the Honeybee (*Apis mellifera*)," *Insect Sociaux* 56, no. 2 (July 2009): 193–201, https://doi.org/10.1007/s00040-009-0012-6. At the same time, experiments show that precocious foragers are also more vulnerable to the risks associated with foraging and tend to have higher mortality rates: Alberto Prado et al., "Honeybee Lifespan: The Critical Role of Pre-Foraging Stage," *Royal Society Open Science* 7 (October 2020): 200998, https://royalsocietypublishing.org/doi/pdf/10.1098/rsos.200998.

5. Known locally as a "bee mother's milk" (*matična mliječ*), the whitish, goopy liquid means much more to the bees than milk does to mammals. Young workers exude it from their mandibles and insert it fresh into cells, one tiny drop at a time, to feed small larvae. The diet of each type of larvae depends on its social sex and hive destiny: drone larvae are more lavishly indulged compared with workers and are weaned later, but the future hive mothers, ripening in long, cone-like wax wombs, are fed the most and most frequently. The milky substance will make them capable of mating shortly after they emerge from the queen cells and keep them reproductive throughout their lives. This royal jelly promotes prodigious early growth, whereby in a matter of several days, queen larvae increase in body weight by a thousand times (translated to human terms, it is as if a five-day-old infant weighed 3.5 tons). The bee queens, or "bee mothers," as they are known in BSC, live up to six years, whereas their female hive kin last from fourteen to forty summer days to up to six winter months.

6. Foragers collect pollen from plant anthers with mouthparts and braid together the pollen dust that sticks to the bees' bodily hair. Pollen grains are mixed up with bees' saliva,

passed onto the forelegs, and deposited into pollen "baskets" (corbiculae) on the hind legs, which weigh around one-tenth of the bees' weight. It takes the bee anywhere from half an hour to four hours to collect the load. On a bee's single leg is a stunning assembly of an estimated hundred thousand to four million individual grains of fine pollen dust, held together by means of static electric charge. Bees deliberately collect an array of pollen, and each type offers distinct bioactive components. Fermented bee bread is even more complex and varied: no two identical samples of bee bread are found in the hive. See, for instance, August Easton-Calabria, Kristian C. Demary, and Nola J. Oner, "Beyond Pollination: Honey Bees (*Apis mellifera*) as Zootherapy Keystone Species," *Frontiers in Ecology and Evolution* 6 (February 2019): 161, https://doi.org/10.3389/fevo.2018.00161; Cédric Alaux et al., "Nutrigenomics in Honey Bees: Digital Gene Expression Analysis of Pollen's Nutritive Effects on Healthy and Varroa-Parasitizes Bees," *BMC Genomics* 12 (October 2011): 496, https://doi.org/10.1186/1471-2164-12-496.

7. John Foster, *After Sustainability: Denial, Hope, Retrieval* (New York: Routledge, 2015), 19. I appreciate many points that Foster's daring book makes while I take issue with his laconic dismissal of contemporary relevance of religion. Not least, as a reader of traditional metaphysics, I can only puzzle over the fact that Foster proposes to treat what he defines as metaphysical problems by other than metaphysical means. As a reader of Islamic metaphysics, I further wonder at the proposition that material or social things are devoid of metaphysical qualities. In any case, while such critique is the staple of scholarly writing, I strived to keep such objections and responses minimal through this book while foregrounding the ethnography and engaging closely, instead, Islamic textual tradition.

8. 42:52.

9. 24:35.

10. For arguments about environmentalism after the end of nature or for ecology without a transcendental ideal of nature, see Bill McKibben, *The End of Nature* (New York: Random House Trade Paperbacks, 2006) and Timothy Morton, *Ecology without Nature: Rethinking Environmental Aesthetics* (Cambridge, MA: Harvard University Press, 2007). The strongest argument for moving ecological politics beyond nature is still Bruno Latour's *Politics of Nature: How to Bring the Sciences into Democracy*, trans. Catherine Porter (Cambridge, MA: Harvard University Press, 2004). Most interesting, however, are the voices in the Euro-American academia that resist the wholesale rejection of the concept of "nature." In the late 1990s, famous anthropologist Marilyn Strathern cautioned that the Western idea of nature is more complex and ambivalent than the nature/culture dichotomy suggested. See Marilyn Strathern, "No Nature, No Culture: The Hagan Case," in *Nature, Culture and Gender*, ed. Carol P. MacCormack and Marilyn Strathern (Cambridge: Cambridge University Press, 1980), 174–222. Philosopher Kate Soper acknowledged all the good reasons that supporters of feminist, queer, and other nonconformist politics object to the normative uses of nature and yet, Soper argues, the idea of nature is indispensable to our thinking and more expansive than it seems in the crude uses. See Kate Soper, *What Is Nature?* (Malden, MA: Blackwell, 1995). Eco-feminist Stacey Alaimo argues that nature—as the domain of material bodies and carnal experiences—was too long bracketed in feminist writing. In the interest of recognizing experiences of environmental injury, however, Alaimo suggests we need a more capacious understanding of the materiality that nature denotes and that extends from green to built to bodily worlds, especially under the conditions of pervasive toxicity. See Stacey Alaimo, *Bodily Natures: Science, Environment, and the Material Stuff* (Bloomington: Indiana University Press, 2010). The concept of nature carries a special valence with in Bosnia and Herzegovina and, arguably, the wider postsocialist Southeast Europe and the former Soviet Union. Ethnographers from the region show that nature, like culture, is a contextual term, strongly connected to popular health concerns, dietary and medical practices, and dacha lifestyles. Latour, *Politics of Nature*; Strathern, "No Nature, No Culture," 174–222; Soper, *What Is Nature?*; Alaimo, *Bodily Natures*; T. J. Demos, *Decolonizing Nature: Contemporary Art and the Politics of Ecology*

(Berlin: Sternberg, 2016). See Melissa Caldwell, *Dacha Idylls: Living Organically in Russia's Countryside* (Berkeley: University of California Press, 2010); Susanna Trnka, *One Blue Child: Asthma, Responsibility, and the Politics of Global Health* (Stanford, CA: Stanford University Press, 2017).

11. Wildness is variously defined as a tendency of novel ecosystems to go wild along unexpected trajectories; as a confusion of categories that takes place in symbiotic or parasitic entanglements; as relationships and dispositions that develop in the contexts of mishaps, as risky and out-of-control dynamics; or as the political place of everyday democratic politics. Jamie Lorimer, *Wildlife in the Anthropocene: Conservation after Nature* (Minneapolis: University of Minnesota Press, 2015); Eben Kirksey, *Emergent Ecologies* (Durham, NC: Duke University Press, 2015); Jane Bennett, *Thoreau's Nature: Ethics, Politics, and the Wild* (Lanham, MD: Rowman & Littlefield, 2002).

12. 41:11.

13. 98:7; 2:25; 7:42; 103:1–3.

14. 31:27.

15. 47:18; Here, in particular, I follow Čolić's translation of the verse. Mustafa Čolić, *Et Tarikatul Muhammedijjetul Islamijjetu: Evidencije i definicije islamskih šerijatskih učenja i vjerovanja* (Visoko: Kaligraf, 1998), 223.

16. Mehmed himself uses herbal insecticides to counter hives' mite infestation, administers herbal prophylactics to counter infection by the microsporidian parasites, and breeds his own queens. He does not treat his fruits and crops with synthetic agrochemicals.

17. Given prized place of royal jelly in the hive, its composition has attracted researchers' interest for more than a century. Spectrometry revealed some 185 organic compounds, including amino acids and hormones. Among the proteins is a family of nine known as "major royal jelly proteins," which are being investigated for their cross-species effects on longevity, fertility, and regeneration. Pasupuleti et al., "Honey, Propolis, and Royal Jelly," 1–21; Derrick C. Wan et al. "Honey Bee Royalactin Unlocks Conserved Pluripotency Pathway in Mammals," *Nature Communications* 9, no. 1 (December 2018): 5078, https://doi.org/10.1038/s41467-018-06256-4. See the UN Food and Agriculture Organization's bulletin on beekeeping products, available online. UN Food and Agriculture Organization, "Chapter 6. Royal Jelly," accessed January 2, 2018, http://www.fao.org/3/w0076e/w0076e16.htm. In addition, potential uses of royal jelly in the treatment of neurodegenerative and aging diseases, diabetes, and cancer are evaluated in an overwhelming number of studies; see, for instance, Mozafar Khazaei, Atefe Ansarian, and Elham Ghanbari, "New Findings on Biological Actions and Clinical Applications of Royal Jelly," *Journal of Dietary Supplements* 15, no. 3 (October 2017): 1–19, https://doi.org/10.1080/19390211.2017.1363843; Pasupuleti et al., "Honey, Propolis, and Royal Jelly," 1–21; Rakesh Kumar Gupta and Stefan Stangaciu, "Apitherapy," in *Beekeeping for Poverty Alleviation and Livelihood Security*, ed. Rakesh Gupta, Wim Reybroeck, Johan W. van Veen, and Anuradha Gupta (Dordrecht: Springer, 2014), 413–446. References to the extensive curative uses of propolis are found in Greco-Roman writings and in medieval and early modern manuscripts on *materia medica* across Eastern Europe, and modern research interest in propolis is notable across Europe, South America, Asia, and the Middle East. See Andrzej Kuropatnicki, Ewelina Szliszka, and Wojciech Kroll, "Historical Aspects of Propolis Research in Modern Times," *Evidence-Based Complementary and Alternative Medicine* 2013 (2013): 964149, http://dx.doi.org/10.1155/2013/964149. Effects of propolis on gram-positive and gram-negative bacteria have been the subject of in vitro and animal studies; see, for instance, Olga Mirzoeva, Ruslan Grishanin, and Philip Calder, "Antimicrobial Action of Propolis and Some of Its Components: The Effects on Growth, Membrane Potential and Motility of Bacteria," *Microbiological Research* 152, no. 3 (September 1997): 239–246, https://doi.org/10.1016/S0944-5013(97)80034-1.

18. At the time of my visit, Dalila was finishing her undergraduate studies in microbiology and writing her thesis on the subject of the inhibiting effects of bee venom on five bacterial cultures, including methicillin-resistant Staphylococcus aureus (MRSA), known as

a "superbug" because of its resistance to many antibiotics. Dalila carried the research project into her master's program.

19. See, for instance, Al-Hafiz Basheer Ahmad Masri, *Animal Welfare in Islam* (Markfield: Islamic Foundation, 2007); Sarra Tlili, *Animals in the Qur'an* (New York: Cambridge University Press, 2012); Fakhar-i-Abbas, *Animals Rights in Islam: Islam and Animal's Rights* (Riga: VDM Verlag, 2009).

20. Ibn al-'Arabi has a fascinating discussion of human dependence on animals and animals' special connection with the divine in the chapter of *Meccan Openings* titled "On Animals" [O Životinjama]. Ibn al-'Arabi, *Mekanska otkrovenja*, 2:588–597.

21. Bediuzzaman Said Nursi, *The Flashes Collection*, trans. Şükran Vahide (Istanbul: Reyan Ofset, 2002), 169.

22. 36:82.

23. Every possible thing craves existence, Ibn al-'Arabi writes. See William Chittick's brilliant discussion of Akbarian ontology in *The Sufi Path of Knowledge*, especially pp. 84–94. Chittick, *The Sufi Path of Knowledge: Ibn al Arabi's Metaphysics of Imagination* (Albany: State University of New York Press, 1989), 84–89.

24. Mustafa Čolić, *Evidencije i definicije islamskih šerijatskih učenja i vjerovanja*, 382.

25. 3:145.

26. 11:103–106.

27. 11:107.

# — 4 —

# Near-End Ecology

## A Devil at Heart

The memory of the ruined apiary still filled Shaykh E. with grief months later. It took him five weeks to get to the hives after the storm had ended. The pockmarked single-lane road to the village in eastern Bosnia was overwhelmed. Creeks spilled out, building up roadblocks of sludge and waste. After the waters receded, whole sections of the road caved in as the asphalt was torn apart by landslide undercurrents. Once he finally made it to the village, he stood at the apiary's foothill. A slope of the hill had cracked and welled up, furrowed like a great turkey wattle. Under the heavy showers that persisted day after day, a landslide was in the making. He turned back and waited some more. A week later, he anxiously climbed the hill. He found drenched bee boxes strewn about the sad orchard, nests empty, and frames with wax comb gone. Someone had robbed his ruined apiary.

Perhaps you can start again? I suggest, voicing my question as a condolence. Shaykh E. shakes his head doubtfully. Starting from scratch is too much of a feat for a hobbyist. Besides, his commitments to the Sufi lodge are growing, he says.

With his well-groomed beard finely peppered with gray hairs, a smart collarless shirt in an eggshell hue, and a wide, outgoing smile,

Shaykh makes a strong impression at our first meeting. His demeanor indicates a former career in public media. He now runs a Sufi lodge with a contemporary flair that makes the spiritual tradition fairly undemanding on the younger urban and semi-urban crowds.

Shaykh E. attracts dervishes, men and women committed to searching for closeness to God. To search for God, one embarks on a path that begins with the greatest hurdle, which is one's own arrogant soul, *nafs*, the self that inclines a human toward evil. A shaykh is a guide, a coach who discerns students' faults and trains them to develop a beautiful character. The lodge also attracts the wider public to weekly gatherings for joint invocations of divine "Beautiful Names." The joint chants arise like the droning of a hive: awake, eager, longing for the nectar flow. Then there is the daily flow of visitors seeking out Shaykh for consultations.

It is July 2014. I have sought out this Sufi elder and teacher (Shaykh is the honorific that carries both meanings) to learn about his beekeeping practice, only to find out that his bees had perished due to disasters that followed the cyclonic storm earlier in May.

These are troubled times, Shaykh says, with ample disorders brewing on the outside and from within. People seek him for help with life and health complaints, he explains, that are evasive to standard psychological and pharmaceutical treatments. Shaykh diagnoses metaphysical ailments. Because humans are not only made of flesh, human suffering and troublemaking also pertain to the *ghayb*, the "hidden," Shaykh clarifies, tapping his chest to indicate the inner core of being.

Shaykh's primary remedies are the Qur'an and honey, two species of Revelation, human and apian. Honey is ingested, the Qur'an is recited, and recitations over honey blend the two potencies together.

*Ghayb*, a loan word from Arabic meaning secret, unseen, and remote, refers to something that needs to be taken on faith. It is not simply inaccessible because it is tucked into some dark corner deep within the folds of flesh and sinew. Rather, Shaykh indicates the reality that cannot be grasped by eyes, ears, or hands but is primary and encompasses the sensuous world that subsists on it and veils us from it. God as pure existence is *ghayb*, while everything that subsists on divine existence at once reveals and hides its source. *Ghayb* also denotes intimacy, as when God says, *Indeed, we have created the human and we know what his soul whispers to him* and *We are nearer to him than the jugular vein.*[1] The implication is that *ghayb* is also too close to see.

This chapter revolves around two extreme weather events: the monster storm of 2014 and the 2017 drought, made more severe by a heat wave dubbed Lucifer. The storm garnered massive international attention and relief and recovery funds while the drought failed to solicit a formal governmental response. Both became landmark events in the international conversations and policy writings about regional climate futures, exemplary of the weather catastrophes that are anticipated to recur and intensify in the Balkans in the coming years. In these accounts, the effects and threats of extreme weather are described technically in terms of aggregate monetary values while demographics guide the determination of differential degrees of local populations' vulnerability to natural hazards. These narratives aim to give a comprehensive sense of weather disasters as ultimately manageable risks to property and assets that can be averted through preparedness and climate-smart investment.

The opening story of a lost apiary, by contrast, brings us to the ground level to show how particular people and their animals fared through extreme weather and the attendant disasters that are regularly worsened by the country's shoddy infrastructure and particularly inept and unwilling environmental governance. Local apiculture affords a particular vantage from which to critically study policy documents on climate disasters and risks precisely because it does not figure as an economically significant sector in Bosnia and Herzegovina (BiH) and because it revolves around things, places, and beings that are highly praised in terms other than monetary. Local beekeepers, a varied but often highly committed group of professionals and hobbyists, are already struggling to keep the apiaries going through the rapidly changing seasons, and extreme weather events frustrate their most resourceful management strategies.

Importantly, this chapter also takes off from that point that Shaykh E. flagged when he tapped his chest arguing that current disorders were raging within as well as around contemporary humans. An ethnographic study does not just expose niches of ecological practice that are economically marginal and consequently underappreciated in the economically focused accounts of disaster risk management.

It also yields a different understanding of the climate change disaster through stories that can hardly be accommodated by policy documents that confidently evaluate the world in trouble—its meaning and worth—and presume to know the means of redeeming it.

Weather disasters and global climate change, along with their implications for honeybees and humans, are storied among devout Bosnian Muslims with recourse to the Revelation. Natural and weather disasters are colloquially glossed as *kijamet*, which is a Slavic transliteration of the Qur'anic Arabic term *al-qiyāmah* with a more expansive, eschatological meaning of the world's end and the trials of judgment that follow. My interlocutors' moods were not apocalyptic in the usual sense of the religious or environmentalist expectations of the nigh end, but the disastrous weather regularly stirred the conversations toward *fasād*, a human-caused disorder, an arrogant mess-making that accelerates the world's race to ruin. The storms of our times blow through the long-winded epoch known in Islamic terms as *ākhīr al-zamān*, the final times.

The common reference for *fasād*, usually translated as corruption, mischief, or evil (in Bosnian-Serbian-Croatian [BSC]: *metež, korupcija*), is the following verse in the Qur'an: *Fasād has appeared in the land and the sea, because of what people's hands have earned, that He may let them taste some consequences of their deeds, and that they may turn back from evil.*[2]

The idea that natural catastrophes have become ample and manifest at present and have been earned by human deeds is the starting point for local conversations about human responsibility for the planetary crisis. Those conversations, by default, extend to contemplations about the human being as when my interlocutors ask, What is the humans' role on Earth, and how did we, as a species, get here, to begin with? These questions serve as welcome pretexts for retelling key Qur'anic narratives, prophetic sayings, and their commentaries to make sense of the present. Telling of the near-end ecology, these stories return to the mythical beginnings to check the tendencies that drive the planet to ruin because the whole point of remembering the Qur'an *is to turn back from evil*, which means, as some translations of the verse above make more obvious, to return to God. At first look, local cosmological stories may sound merely fantastic or theistic and, therefore, incomparable with the sober prose of operational documents that assess the tolls of a past disaster or anticipate future risks and, moreover, irrelevant to the real and pressing challenge of grasping the implications of climate change.

Contemporary social thinkers, however, insist that the current scale and intensity of global environmental and climate crisis necessitate that we ask again: What is the human species like? What is the meaning of the human in the light of its evident capacity to drive its home planet to collapse. The answers reveal nothing short of late-modern human soul-searching.[3]

Policy documents written in the wake or in anticipation of extreme weather catastrophes in the Balkan region are themselves and despite themselves narratives of a near-end ecology. They take catastrophic weather as a fact of climate science and as an urgent pretext to define the key values at stake and the means of protecting them through the precarious times they anticipate. Steering away from questions of human responsibility, they project a bright future on account of a presentist cosmology.

Thinking through the *fasād*, Bosnian Muslim beekeepers and bee lovers regularly compare humans with honeybees and their respective relationships with the world and God. Unlike bees, humans are drawn to bonding with Shaitan, the devil, which achieves divine distance, another word for divine wrath. This is where the story gets complicated, and rather than getting ahead of myself, I will let several local storytellers untangle the fine threads of the cosmology that, with hand on chest, implicates each embodied being in responsibility for the world's ruination, embroils our hearts' beats with the crises blasting across the lands and the seas.

## *Hives in the Storm*

No one knows how many hives were carried off by the floods or ruined in the aftermath of the 2014 storm. Beekeeping in BiH is a small agricultural sector in which most apiculturists are self-employed, off the records, while the small incomes they make—along with investments and losses—are rarely formally registered as such. Beekeepers, a well-networked, if diffused, professional community, estimated losses in thousands of hives. The story of Shaykh E.'s lost apiary gives an idea of the predicaments suffered by the beekeepers who were directly affected by the floods.

The cyclone that hit the country in May was extraordinary. Three months' worth of rain fell within three days.[4] The incoming storm gained force on its course through Bosnia's mountains while the soil and the rivers had already been inundated by the snow thaw and heavy spring rains. The watershed in the lowlands swelled with volumes of water, timber, and stones that rushed in with the streams from the highlands.

The flooding was most catastrophic in the country's northeast. The Bosna River, which flows northward through the country's densely populated urban and industrial area, swept along mounds of environmental waste, casually deposited along its banks from meat, leather, steel, and cellulose factories, among others, and fed its toxic trail into the Sava, an international river that joins the Danube farther northeast in Belgrade, the Serbian capital. Other tributaries of the Sava hurled in waste and water from the rivers of eastern Bosnia, which are likewise massively polluted by the industries that run on coal and dispose of emetic loads of sludge and heavy metals to riverbeds. The Sava soared by tenfold of the average rate at which this flood-prone river reaches alarm thresholds.

Before the storm, the country's mobile beekeepers traveled to the Sava River basin, known as Posavina, attracted, as every year before, by the prospects of nectar flow. The temperate riverine climate and expanses of black locust forest and false indigo thickets make this region the early point of departure for the annual cross-country honey hunt. The hives were parked in the fields fringed by black locust trees budding in the eye of the storm.

When the storm clouds blew past, these honey forage fronts were left far within the disaster zone. International relief experts, state agencies, and volunteers from across the country rushed to the region to help evacuate and assist the nearly one million people whose homes were submerged and whose water sources, crops, and food stocks were contaminated. In the wake of receding water, erosions and landslides followed while unexploded ordnances and land mines left over from the 1990s war were uprooted and widely replanted, randomly arming fields and trails.

Floods and landslides impeded traffic along the poorly maintained two-lane trunk roads that course through the region, while the small roads and earthen paths that beekeepers travel to seek out forage were rendered inaccessible. Hives were left in the swamped fields with the forage prospects ruined and beekeepers unable to evacuate or assist honeybees with emergency food.

That the beekeepers' losses and predicaments were generally unrecognized was the staple complaint within the local apicultural community. An article published in a beekeeping magazine by an apiarist and biology high school professor expressed general grievances: "In all normal countries, the state ministries respond to situations like this with aid resources, while in our state they do not. Unfortunately, even [our] loud statements about the importance of bees, first of all as pollinators, were left unheard."[5]

To secure governmental attention, beekeepers emphasized the bees' wider pollination eco-service, couching their arguments in environmental and secular terms of the state's formal discourse, which nonetheless failed to make an impression. Lack of financial assistance aside, the author complained about more general inconsideration shown by local levels of government toward the beekeepers and bees. A gross example was the aerial insecticide spraying conducted across the country, which targeted the mosquito populations that surged after the floods and inadvertently hurt the honeybees.

The author also made it clear that beekeepers' concerns and losses extend beyond the storm and floods. The disaster, beekeepers commonly suggested, was an enduring event. Black locust forage was ruined, and the weather conditions remained worrisome: "We should ask ourselves: what if the rains and low temperatures persist and, God forbid, there is no chestnut, linden, or meadow flow?"[6]

The rest of the foraging season was as bleak as the author worried it might be. The weather did not go back to normal. Nor did the bad year begin with the May storm.

The year 2014 had been strange from the very beginning, beekeepers noted. Temperatures in mid-January were unseasonably high. High enough for early spring plants to bloom and offer up nectar and pollen, which in turn stimulated the "bee mothers" (queens) to lay eggs. The new brood in turn energized foragers' field efforts, and the hives quickened with the expansive insect enthusiasm that grows community numbers, secures food supplies, and puts bees into the mood for swarming.

The hives' spring expansion is typically supported by the saved stocks of the preceding year's food, which are used sparingly over relatively dormant winters. They are then opened to feed the fully awakened insect appetites and fuel the tremendous effort of nurturing new bees and building wax. Ideally, the supplies would be replenished, first by the shy but steady nectar and pollen incomes from the early bloomers, then more substantially with fruit blossoms and, finally, with the windfall of strong honey flow by the late spring flora, such as black locust.

The trouble with false springs, which are unseasonably warm spells in late winter, is that they prematurely send cues to bees and their companion plants to embark on spring development whereas winter conditions are bound to set back in. And indeed, come February, temperatures dropped below freezing and honeybees withdrew into clusters, which is how the collective weathers the winters: the outer ring of bees shudders to keep the inner core warm. They take turns on the cold fronts. The prematurely expanded brood falls outside the cluster and so remains unheated. Suboptimal temperatures adversely affect the broods' development with lasting consequences for the bees that hatch.[7]

If the nest expansion spent the hives' honey supplies, the temperature drop locked the bees in without the fuel necessary to support the task of the cluster's warming. Consequently, starvation became a real threat until warmer weather provided greater mobility and forage opportunities. Introducing emergency food at that point makes no difference because the clustering bees can neither reach nor process it.

Fruit trees bloomed next throughout the persistent rains in late spring, and the bees remained homebound. The black locust blossomed during the cyclone, and heavy showers persisted throughout the summer into the fall.

As one apiarist, Pašezad, puts it: "A beekeeper must buy sugar to prevent the bees from starving, and then his neighbor sees him and declares, loud enough for others to hear: 'Nts. Nts. Nts.' There goes the beekeeper, making 'honey'!" This joke that Pašezad tells at apiculturists' expense expresses deep local misgivings about the sugar in the hives.

Local beekeepers claim that honeybees overwinter poorly on exclusively sugar-based food supplies. Honey is a nutritionally rich and complex substance that besides carbohydrates contains a wide array of minerals, vitamins, enzymes, as well as traces of pollen, which itself is a rich substance. To provide the bees with essential nutrients and a prophylactic diet, local beekeepers advise preparing the bees for the winter with half or at least a third of food stores composed of "genuine honey."

A bad year makes honey scarce and inflates suspicions about the authenticity of the honey offered for sale. Considering that honey is praised as a product of divine revelation, a nutritional supplement, a prophylactic, and a folk remedy, its purity is of high concern to Bosnian Muslims. It is a public secret that several honey-counterfeiting facilities operate in the country, while their technologies for simulating the appearance of honeybees' honey, the beekeepers report, are notably advancing. Moreover, sugar that is fed to the bees is processed into honey, which, once harvested and short of laboratory analysis, does not betray its nonfloral origins. To the onlooking neighbors, there is hardly any difference between the emergency feeding of bees through dearth and deliberate attempts to produce sugar-based, fake honey for sale.

As another article published in *BiH Beekeeper* magazine suggested, the year 2014 disproved the local beekeepers' traditional wisdom that exceptional nectar flow happens every four years, making up for one poor year and two mediocre years in the meantime. According to that count, 2014 was supposed to be a "golden year." "Why is this happening to us? Weren't there too many bad things already? Where is this leading to?" the author asks and ends by simply admitting: "Hard questions!"[8]

## *Honeybees, Humans, and Other Worlds*

Hard questions were regularly raised in personal conversations about bees in the summer of 2014, while the contemplations went far, searching for the causes and meanings of the current catastrophe.

"Just look at the ecological disasters around us," Hafiz Ahmed tells me. "What a *fasād* on earth we have made." Over Turkish coffee in his office in mid-June, the tall, tranquil young man speaks in the language that casually combines concerns of ecology and the insights of the

Revelation. "The global disappearance of the bees is the clear sign of the world's undoing," the Hafiz says.

Hafiz, which means "safekeeper," is an honorific for the one who has memorized the Qur'an and who recites it regularly to keep the Book from being forgotten. Hafiz Ahmed is also a novice beekeeper, although his recent appointment as the main imam with the Islamic Community's office in a northeastern Bosnian town has made him too busy for the bees.[9] Just for the time being, he adds cheerfully. Inshallah, God willing, he will go back to beekeeping. Hafiz Ahmed recalls how difficult it was to memorize the Qur'anic chapter titled "The Bee." While he studied it, the rhythm of its sounds struck him as particularly complex, and the calligraphy on its pages appeared to him as vigorous as a swarm.

He recites for me the verses of the "The Bee" chapter and gives a commentary: "We translate it as: 'Your Lord revealed to the bee.' The verb used in Arabic is *ewḥā*, whose root letters make up the noun *waḥy*, 'the Revelation,' the same term which is used for the revelations received by the human prophets. The bee is given instructions on how to live and what to do. It lives by the Revelation it received while people often do not. If we followed the Revelation as wholeheartedly as the bees, there'll be no *fasād* on Earth. Bees incline toward perfect harmony, toward keeping things together, within hives and outside."

"What about other animals?" I ask.

Honeybees are special, the Hafiz says, then adds: "The way I see it, all creatures, actually, have a revelation of sorts and live by it. It's what is meant by an 'instinct.' But the human is something else. It has the soul and a chance to disobey and err. This is why we need the Revelation, to discern the right from the wrong."

Pausing to think, Ahmed says: "The honeybee is a symbol of beauty, purity, and striving in Islam. We Muslims ought to incline to be like bees." It is easier said than done, he adjoins, observing that ecological values that are core to Islam are seldom appreciated. "Everything is, obviously, placed at human disposal. But that is why, precisely, we shouldn't mess it with it. We see clear signs that things are off. We are altering the natural conditions under which bees and animals live, by polluting the air, the water, with our technologies we bring out new diseases . . . The human species will ruin itself. By running after profits, after whatever it is that leads them astray."

According to Islamic sources, the human soul is a subtle reality that animates the body but occupies the metaphysical heart. The heart is that which constantly stirs and turns, from devotion to treachery, from hope to despair. The heart wavers between the *nafs*, the self-centered soul that aspires to be a god and tolerates no guidance, and the *rūh*, the spirit longing for God. Many things move the heart, which by its nature is already unstable, and receptive to influences. The heart is always weathering storms of conflicting thoughts and desires.

While descent of divine Revelation, in the sense of the Scriptures has ended with the Qur'an, the Hafiz says, a related state of inspiration is something that people experience all the time, and its nature depends on the condition of the human's soul. "The inspiration can be *ruhani* or *shaitani*," he adds. In other words, it can speak to and from the spirit, or it can arrive with shaitan, *who whispers in the human breasts*.[10]

"Honeybees are interesting, but they are only one of many worlds," says Shaykh S., who has politely agreed to see me at Hafiz Ahmed's recommendation. A son of a beekeeper, the soft-spoken, reserved, silver-bearded Sufi holds an appointment in the Islamic legal affairs office and, less publicly, occupies the post of spiritual elder or teacher. "This is why God is called the Lord of the Worlds. Worlds are different, but the God is One, which is why all the worlds, the visible and the invisible ones, are connected."

Shaykh S. carries on: "Muslims believe in the world of angels and the jinn; those are worlds parallel to ours. We are limited to the visible reality, but all these other worlds are there, though we don't perceive them. . . . Our own limitations are not the limitations on the reality itself. Our vision doesn't go past our eyesight, but that doesn't mean there's nothing else out there. Our eyes hit their limits, that's all."

Faith in *ghayb* is mandatory for Muslims, *who believe in the Unseen [ghayb] and perform the prayer and give of that we have provided them*.[11] Taking God at His word, this verse implies, presumes faith in what one cannot verify, and the faith is confirmed practically through compliance with commands and prohibitions aimed at the selfish *nafs*. Throughout, the Book couples faith with concrete acts—such as fasting, giving charity, weighing goods justly, keeping hope, and such, emphasizing choice: *Whosoever does good, does so for his own soul's gain, and whosoever does evil, it is to his own loss. Your Lord does not wrong His subjects.*[12]

"It is not the bees who cause disorder." Shaykh S. continues his outline of Islamic cosmology. "The human being alone can." The jinn, too, according to the Qur'an, are prone to disorder, but the Shaykh stays focused on the human. Unlike other beings in the earthly world, the "human has the faculty of reasoning and relative freedom of choice, it can do good or make a mess. It will be held responsible," Shaykh explains, implying an accountability that the Qur'an insists extends to the smallest details. *And We shall set up the just balances for the Resurrection Day so that not one soul shall be wronged in the least: be the deed the weight of the mustard seed, we shall bring it forth. And We suffice as an Accountant.*[13]

Ultimately, humans will be responsible for the world's collapse, which Shaykh says is an event that may already be underway: "We usually imagine a single closing event—The End—but for all we know, the ending could be a slow process by which the worlds are vanishing, one by one."

### The Devil's Prayer

"The natural order everywhere is being disrupted," Adil reflects. Retired from his post as imam in a city mosque, Adil runs a branch of a Bosnian Muslim charity organization, *Merhamet*. In June 2014, the whole office was busy with humanitarian projects aiding the population affected by the floods and landslides.

"Bees are extremely important for the ecological system since they pollinate so many plants. And bees make medicine. But modern humans, who give priority to monetary values, have endangered the bees. There's too much chemistry applied to eradicate insects, too much poison sprayed around. Mosquitos are now air-sprayed without us [the beekeepers] being forewarned. An apple tree is regularly treated with herbicides twenty to thirty times a year! That inevitably causes havoc in an ecology. A honeybee lands on the leaf and perishes. And without them, things can grow, but they don't give a yield. The Divine laws in nature are being disrupted, whereas the Dear God orders a respect for nature. And when the bees disappear, the people too will disappear."

Adil delivers the point matter-of-factly and pauses to pour more coffee into delicate white finjans.

In the years to come, Imam Adil will become one of my mentors in beekeeping and gardening, especially once his retirement affords him full-time residency on his paternal land. My sister and I will spend time at his village apiary following him in practical tasks—grafting, swarm catching—admiring the latest ingenious apiary devices that he has invented—a bee smoker out of a recycled can that keeps the smoke steady, for one—as well as his resourcefulness—his work pants' pockets are stuffed with an array of essential tools. We do not hide being impressed, and Adil graciously accepts the compliments with bouts of deep-throated laughter.

In the summer of 2014, Adil's mood is more sober. Dressed simply but well in a fine woolen pullover and a tipped black beret, typical of an older generation of urban Bosnian Muslims, Adil is a storyteller animated by spurts of bright energy.

"By working, an *insan* [a loan word from the Arabic *insān*, a human] is supposed to contribute to the natural order. Whenever he fails to do so, nature responds with a slap. Someone says to himself: 'Too many birds around here, I'll get rid of them,' but then locusts come or caterpillars multiply and ruin everything. So it goes, whenever humans act on their whim. And of all the beings on Earth, the human is the most accountable. The human should really act as a *khalifah* [Arabic for a successor, representative, or vicegerent] and contribute to the world in that role. After all, that's what God had appointed him to do. But *khalifah* is devout, a man of faith and a man of faith is a good-doer. Faith without good deeds is like a tree that bears no fruits."[14]

"How is the human appointed to this role?" I ask, and Imam Adil indulges my curiosity.

"God created nature—earth, skies, and so on. Everything that we know and so much in cosmos that is *ghayb*, unknown to us. Earth was brought about, and time passed. Then, one day, God asks the angels and the jinn, the sun and the moon, He asks the mountains: 'Will you take up the responsibility for the earth, the *khīlafah*?' Everyone politely declines, except for the human. Our kind is hasty. Since then, Adam and his progeny carry on the *khīlafah* on earth. Everything bids them service, for the love of God. But as a *khalifah*, the human is the most responsible one: to ensure that the natural laws are respected and that the balance isn't disturbed. But you see how that goes. Pollution, disruption..."

"And the beekeeping? How does it relate to being *khālifah* or being devout?" I intercept. Instead of replying, Imam Adil startles me with a counterquestion: "Why is Shaitan [Satan] a sworn enemy to the human?" Without waiting for me to answer, he carries on with the story of human origins.

"Once Adam took up the post, God invites all the heavens' residents to honor the human with a *sajdah* [prostration]. The angels are incredulous—'You've appointed the species that is prone to *fasād*, messmaking? While we adore you so faithfully,' they said, feeling slighted. Still, everyone complies, except Shaitan by the name of Iblīs, who was a respected jinn. So devout he was at the time, that he earned pride of place among the angels.

'Why should I honor him whom You've created from lowly earth? And I am made of a smokeless flame,' Iblīs says proudly and then points out: 'This kind would kill his own brother! He should honor me, instead.' He argues and argues, clever as he is, but the point he missed is that it was God's order that he scoffed at.

'Very well,' God says in turn, 'you earned yourself the Fire [*Jahannam*].' Iblīs bites his lip. Then he says: 'Have I not been a good subject of yours thus far? Do I not deserve to have a prayer heard? So, here, I implore You: defer my exile until the end of time, so I can bring to ruin all who are willing to come along with me.' And God grants Shaitan's wish. Then He swears: 'But you will never sway my sincere subjects.'

And so, from then on, Shaitan will not leave humans at peace. If someone wants to do some good or perform the prostration in prayer, Shaitan strives to distract him. . . . He is busy, working to ruin the world."

The look on my face must tell Imam Adil that I am at a loss, still waiting for the founding myth to deliver an explicitly ecological or apicultural point.

Imam Adil kindly clarifies: "Ever since, God has been sending prophets to the humans to guide them. Faith is an intrinsic part of life and of the order in nature. Everything I do, everything about my life relates to my faith (*imān*). I'm striving not to be swayed by Shaitan."

## A Jinn Epidemic

As Shaykh S. has said, the faith in *ghayb* is incumbent on practicing Muslims. Angels, jinn, and Shaitan are invoked in the Qur'an not as metaphors but as resident beings in a cosmos composed of many worlds as well as characters involved in a range of quotidian relationships with humans.

At the same time, in BiH, rationalist philosophy and scientific naturalism practically define medical, legal, and scientific realities of the modern state and exercise a powerful hold on commonsense notions of what is real and possible. Institutions and ideologies of scientific socialism waged a war on "folk superstitions," pitting religious rites against the tenets of modern life and secular science. Bosnian Muslim Islamic tradition, given its Ottoman and Oriental history, was perceived as particularly embarrassing to socialist Yugoslav aspirations to, essentially, a Western modernity.

After Yugoslav socialism fell apart in BiH, as a political economy and a cultural project of forging secular, socialist identities that were ethnically mute, some modern Bosnian Muslims underwent religious reeducation. Since the 1990s, Islam had been returning to public life, coloring the forms of Muslim ethnic identity and political affiliation. In a quieter manner, Islam was undergoing rediscovery and a reevaluation as a private resource for thinking about the world and caring for the self.

Shaykh E., the Sufi elder who lost his apiary to the floods, is one of many experts whom people seek out with life and health issues that cannot be grasped by biomedical diagnostics or rational explanations. Patients complain that places, homes, and objects have turned threatening and unstable; their bodies and skins have been exposed to deeply disturbing influences; their feelings, words, and acts have become strange, savage, destructive, and out of character. Consultations entail taking a physical examination and a detailed health history. The ultimate test, however, is the patient's response to the therapist's recitation of the Qur'an. A strong, allergic reaction to the Qur'anic verses is taken as an indication of jinn-meddling influences.

Otherwise, jinn are subtle beings. Unless they interfere, they live side by side with humans and animals, as unnoticeable as the microbes that pervade our environment. The combination of j-n-n consonants in the Arabic root word connotes veiling. Jinn are hiding in plain view: think of a garden (*jannah*) where light plays hide-and-seek through the trees' canopies. Think of a fetus (*jinnīn*) enveloped in multiple membranes beneath the skin of an expanded belly. Or think of madness (*junūn*) that is said to veil reason. The word's etymology relates jinn to familiar things and so helps flesh out the ordinariness of their presence, despite their occult reputation.

Although intangible, jinn are a worldly species. Corporeal, mortal, and gendered, they hunger, drink, and couple, lust and long, err and ache. Animal-like in their carnal lives and kindred to humans by virtue of their willful souls, jinn bodies are composed of an element finer than human clay yet coarser than the light that makes up the angels.

Moreover, jinn are environmental agents. "They feed on phosphorous," Sufi elder Shaykh Ayne tells me, providing a clue to the jinn's worldwide range. Phosphorous is a mineral element ubiquitous on our planet. Variously bound with oxygen, it builds phosphate, an essential nutrient for all living organisms. In staple foods—meat, eggs, dairy, whole grains, and potatoes—and the fizz of soda drinks, it courses through our diets, growing and repairing tissues and cells. It builds mammalian teeth and bones. The phosphate-rich human urine joins the wider environmental phosphorus cycle that moves the element from the kidneys to the soil, rivers, and oceans. Modern agriculture and hygiene habits flush excessive amounts of phosphorus with fertilizers and detergents into waterways. In short, the jinn's diet makes them our close companions in the biosphere while moderns conspicuously generate a food surplus of the fiery species' essential nutrients.[15]

While jinn live in communities that are as multicultural and religiously plural as are human societies, their igneous nature makes the species temperamental, arrogant, "and intellectually unstable."[16] All jinn feed on phosphorous, but some thrive on energies generated through human rupture, wreckage, and waste. Jinn are affective parasites; they nourish themselves on the discharges of the *nafs* gone wild with toxic feelings of grief, greed, or rage.

Whereas traditionally jinn were said to reside on lonely and lowly grounds, in graveyards and garbage dumps, in sewage and rubble, late-modern industrial and consumer culture has widened the jinn range.

In contemporary BiH, communal and toxic waste is deposited in neighborhoods and environmentally unregulated landfills, mixed into asphalt, industrial toxins pave the regional highways, fed into watershed and illegally incinerated, they saturate the very air.[17] Floods and landslides ruin homes and gardens, disturb former frontlines, wash out entire cemeteries, and destroy livelihoods and savings. Disasters bring humans uncomfortably and irresistibly close, and the jinn negotiate the interspecies contact with ambivalence.

Shaykh E. treats the current surge in jinn disorders among his patients caused by fatal encounters between the two species. The length, frequency, and particulars of the treatment are established on a case-by-case basis.

Shaykh E., however, insists that his patients undertake a home treatment as well. If they protest that they are incompetent, Shaykh says: "I tell them: you can do it! It's not me healing you anyhow. I cannot heal you; I personally have no such powers. God heals you. I can only plead on your behalf, but, really, this issue is between you and God—you ought to carry on with praying (*trebaš učit*). God doesn't want your money, he doesn't need any gifts, but he wants that stiff neck of yours to bow down," Shaykh says excitedly, with a slap to his forehead.

His gesture signals a *sajdah*. The position of prostration with one's forehead to the ground is a key moment in the daily ritual prayer (*salat*) and an utterly embodied act of surrender, which the devil refused to perform at God's bidding. The position humiliates *nafs* and affords to *rūh* a daily audience with God from which to seek divine closeness and protection.[18] More broadly, his choice of words (*učiti*) recommends the recitation of the Qur'an in and outside the ritual prayer. Shaykh E., in short, is reminding his patients to observe the prescribed prayers as a trusted means of mending the distance between humans and God, the vector of jinn infection.

Honey, the other product of divine revelation, is said to be the only foodstuff that the jinn and Shaitan cannot contaminate. As is the case with honey used in the contemporary pharmaceutical industry and posttraumatic care, medical-grade honey is defined by its purity: it is uncontaminated by xenobiotics and undiluted by sugar. With prayers recited over it, honey is prescribed as doubly efficacious, and Shaykh E.'s patients consume it for the length of the treatment. With his apiary ruined, Shaykh urges patients to obtain their own supplies of "genuine honey" (*pravi med*).

## *Etiology of Fasād*

The jinn epidemic aside, metaphysical ailments that preoccupy devout Bosnian Muslims—and Sufis and their guides, in particular—concern more prevalent, intimate, and altogether insistent interferences. Known as disorders of the metaphysical heart or the metaphysical *insān*, these are dispositions and desires that string together a human heart and no one other than Shaitan.

Iblīs, whom Imam Adil's story described as conceited and clever, enraged by the show of divine affection toward humans, is the original Shaitan, but the name now denotes many, both jinn and human, because the devil is not a species but a membership in a professional club.

Beekeeper Mustafa, an old acquaintance whom I meet at his small kiosk in town one day in the summer of 2014, tells me: "When Iblīs saw how beautiful was Adam, peace be on him, he got so jealous, so jealous! He was about to burst with envy. But then he noticed that Adam was hollow, and, overjoyed, Iblīs held his peace."

"Hollow?" I asked.

The hour was slow, and Mustafa, waiting for customers, was passing the time while reading the Qur'an. "Yup," he replies, "hollow. With bodily openings through which Shaitan can come in. He moves through the human body, with blood."

The devil has always inspired a wide range of contemplations in Islamic sources. A dear friend, Zejd, a dervish groomed at the side of a luminary Bosnian Shaykh, suggests a more expansive image.

"Shaitan courses with blood," Zejd writes to me. "It's the virus. It's the thought. It is what day is to night. God creates the night and couples it with the day. *And of everything, we created pairs.*[19] God brings about an angel that inspires beautiful thoughts, from the right-hand side and God gives the left, the side of Shaitan. He makes everything in pairs and He remains single, uncoupled, peerless. This all brings us back to the beginning, since before the time."

The human quarrel with Shaitan and Shaitan's pact with God, Zejd implies, make up the primordial, offsetting tension within binary complementaries that underly the cosmos and the nature of humans. When Shaitan is defined as so fluid and restless—coursing, mutating—and so involved with one's person—it circles through the veins, it whispers to the human heart—discerning its presence or doings seems like a feat. At all times, Shaitan is coupled with *nafs*, the soul self, and this coupling inclines the heart back and forth, from left to right, from good to evil, from day to night.

The Qur'an calls Shaitan a "manifest enemy" and warns people not to follow in Shaitan's footsteps. That which is manifest is sometimes hardest to discern, save for its traces. Footsteps are what you follow on the trail of someone who is leading but also evading. *O you who believe, do not follow the footsteps of Shaitan. And whoever follows the footsteps of Shaitan, surely, he bids to indecency and evil.*[20] An intangible being known by its traces, Shaitan compels and seduces, reasons and advises, flatters and distracts with deceiving promises, all while flaring up feelings and dispositions that spoil the human heart.

The devil leads astray those who believe—to whom the verses are addressed—as well as *whosoever follows*. Because faith alone, implicitly, is not enough to tell or resist Shaitan, its insidious influence requires investigation and self-inquiry. Likewise, a belief is not the condition for the devil's company or his appeal.

It takes an insight to tell a devil, Sufis suggest. An experienced guide helps one to get to know one's self, God, and Shaitan, and tell them apart. This knowledge is exercised in a day-to-day effort to cultivate a virtuous heart, a heart fit for meeting God.

Working on one's heart is, ideally, an everyday Muslim project. The Qur'an suggests as much: *On the Day when neither wealth nor sons will avail save for one who comes to God with a pure heart.*[21] Purification of the heart presumes cleansing or growing of the soul: *Indeed, fortunate is the one who purifies it. And, indeed, fails the one who obscures it.*[22]

Fulfilling ritual obligations purifies the heart. Humbleness and sincerity are prerequisites, so prostration becomes the threshold from which to begin. If the heart bows gladly, the inner journey onward to the divine vicinity begins, even before death and the world's end usher in the day of standing before God.

Sufis who are Muslims devoted to lifelong "polishing of their hearts," as the metaphor frequently goes, have written discourses and practical manuals on the subject of metaphysical illnesses and their treatments throughout history. One cherished reference in a close circle of readers in BiH drawn to Sufism is the book titled the *Heart's Health and Illnesses of Metaphysical Insān*.

A translation and commentary of the sixteenth-century work by Imam Birgivi by a remarkable thinker, Bosnian Sufi Shaykh Mustafa Čolić, the book is a veritable diagnostic and treatment manual for sixty metaphysical ailments.

Among them is *jahl*, ignorance, which can be either simple and treatable with education that stirs up heart's curiosity, or complex, as in the of case of the arrogant folk who take their single-mindedness as a license for passing judgment on the God and others in the world. Other diagnostic terms include *ujub*, self-satisfaction, and the closely related *kibur*, arrogance. Shaykh Čolić describes self-satisfaction as a case of smugly mistaking divine bestowals for one's own virtues and gains. Arrogance is assuming greatness, which rightfully belongs to God alone. Metaphysical illnesses arise from an array of interconnected causes and exhibit overlapping symptoms—arrogance is a form of ignorance, for instance. The bottom line is that treating metaphysical illnesses requires changing a way of life because the self is inseparable from relationships to the outward world. Fixing or worrying about the world at large starts by work first on the troublesome self.

Implicitly, Shaitan is an open enemy because he is aware of his subject relationship to God. Shaitan has never denied God: *Indeed, I fear God, the Lord of the Worlds*, says the devil in the verses of the Qur'an.[23] He refused to obey the divine command out of jealousy and a devotion that adored his own self-regard.

Shaitan is a clear enemy also because he has openly declared his intent to lead to ruin, and his knowledge of God presumes that he knows the difference between right and wrong, bliss and ruin. He intentionally misleads humans—and the jinn—and attracts them with promises he cannot fulfill. Once all is said and done, Shaitan *will say: "God surely gave you a true promise and I promised you then failed you. I had no power over you, except that I invited you and you responded. So do not blame me, but blame yourselves; I cannot help you nor can you help me. I have nothing to do with the fact that you were associating me with God. As for the evildoers, there is the painful doom."*[24]

Fig. 4.1 Heart medicine

Contemplating the devil, Zejd writes: "If Shaitan is a manifest enemy, what, then, are the secret enemies like?" Doing *fasād*, the Qur'an suggests, may be deliberate, deceitful, and wrongheaded but oblivious to the harm done: *When they are told "Do not make fasād on earth," they reply "But why, we are only peace-making."*[25] The secret enemies make of the mess a sound reason, a remedy for setting things right.

### *Lucifer 2017*

In August 2017, a heat wave of blistering intensity engulfed Southern Europe. Temperatures soared from 38°C to 42°C (100°F–108°F). Dubbed "Lucifer" in the international media reports, the heat wave branded the extreme weather event with an infernal reference.

Lucifer worsened the already grave effects of the summer-long drought that gripped BiH along with the wider Balkan region and Mediterranean Europe. The heat wave further evaporated moisture from the ground surface, and parched soil and withered vegetation suppressed the chances of rainfall. With high temperatures persisting into the night, there was no respite for plants, animals, or people, and the heat stress built up across species. Consequences rippled out, and the countryside bore the brunt of the heat's burden.[26]

Fruit and produce farmers reported from their scorched fields: pastures yielded a single harvest where three had been expected. Trees and crops never bore fruit, or they shed the fruit unripe. Dairy cattle reduced milk yield, declined suckling calves, and abstained from feeding. Water sources in villages dried. Trees wilted and wildfires spread, especially across Herzegovina, the country's Mediterranean south. The state's firefighting response was typically underfunded and underequipped as well as hampered by the political blackmailing that regularly takes place across the two ethnically divided and belligerent administrative entities of the Bosnian state. As the wildfires raged, the federal government initiated a proposal to revoke firefighters' hazard pay just as occupational risks were soaring, treating the climate-related emergency with a brazen disregard.

Farmers across the country had been calling for BiH entity governments to declare a state of natural disaster since April 2017 when late frost and snowstorms damaged the crops. As the year progressed and extreme heat set in, calls for formal declarations of an emergency multiplied. All were declined.[27]

Droughts are rather discreet weather catastrophes and, in the absence of wildfires or famines, do not broadcast as dramatic threats to lives or property. Although it was region-wide and severe, the drought was ignored by the local government. At the time, public discussions of extreme weather and global climate change were practically absent in BiH, and the drought with Lucifer in tow was described as an age-old misfortune, a rare and passing phenomenon. The government's inattentiveness, however, was not due to climate innocence. Rather, it was born of a postwar style of governance that doggedly narrowed the scope of domestic political concerns to ethno-national electoral interests to which climate and environment seem particularly irrelevant.

Since 2005, formal assessments of the country's risk and vulnerabilities were periodically conducted by experts from the bipartite Bosnian state. These assessments highlighted droughts and floods as the country's major future threats. They acknowledged that precipitation regimes have changed since the turn of the millennium and that prolonged dry periods alternated with events of severe rainfall. Because of its topography and poor infrastructure, the country was described as especially flood prone.

Massive flooding in 2010 had proved the state's river management strategies to be seriously inadequate and its cross-border coordination politically hampered, providing cautionary lessons and blueprints for subsequent governmental assessments of disaster risks.[28] These routine assessments, however, had few practical implications, as some of the experts consulted in their drafting had complained of.

The storm of May 2014, however, became the watershed event for official accounts of the country's climate future. The dramatic scale of flooding precipitated a high-profile international humanitarian response, and the demands for cleanup and recovery solicited offers of foreign loans and aid funds. By engaging the attention of the international relief and development industry, catastrophic weather events swiftly made climate change effects a concern of national governance.

In the weeks after the flooding, the BiH government, with technical and financial assistance from the European Union, UN agencies, and the World Bank, conducted an assessment of recovery needs. Published in anticipation of the upcoming International Donors' Conference, the "Recovery Needs Assessment" framed the storm as an extreme weather event associated with global climate change.[29] Securing a pledge of €809.2 million in recovery funds, the BiH government formally agreed to implement flood prevention and flood risk management and to invest in a climate-resilient infrastructure.

The "Assessment," moreover, predicted catastrophic futures: "The severity of extreme events like droughts, heat waves, forest fires and flooding has intensified over the last few decades. This trend is expected to accelerate in the future as a result of climate change, leading, together with changes in land-use patterns and increased human settlements in areas that are prone to disasters to increased hydro-meteorological and climate-related risks in the coming years."[30]

A disastrous future notwithstanding, financial opportunities fostered by disaster management were duly recognized: "It must be taken into consideration," the "Assessment" reads, "that the disaster can create new possibility for prosperity through job creation programmes that could help jumpstart and expand growth through the recovery process and reconstruction investment."[31]

A focus on disaster management renders extreme weather a purely economic matter. Climate change itself is redefined not as the mounting catastrophe with unknown consequences for life on Earth but as an emergent landscape of risks and opportunities. For as long as economic losses are minimized or recompensed, the climate future can be anything. Except apocalyptic. Economy bears the language of crisis, collapse, and depression but not the tone of warnings that intimates finite values or a closed horizon of economic development.

## Seek Refuge from the Devil

"The region's future, according to the latest World Bank projections, is bright." So begins an opinion statement by the World Bank's director for the Western Balkans, issued in 2018. The optimistic projections derive from the indicators of economic growth, such as rates of employment, higher wages, and improved living standards.

The director's statement continues on a cautionary note. "However, there are clouds gathering on the horizon. In parallel to this welcome growth, climate and disaster risks are also increasing, putting vulnerable communities in jeopardy."[32] The 2014 drought and the 2017 drought accompanied by the heat wave are cited as proof of "how vulnerable the region is to climate-related shocks."[33] The director's statement goes on to say that "weather extremes like these are fast becoming the new normal—by the end of the century, in fact, average regional temperatures could rise by as much as 4 degrees Celsius or more above pre-industrial levels. This would mean more frequent droughts, reduced agricultural production, severe water shortages, and less hydropower energy—threatening to disrupt decades of important development gains. This is simply too high a cost."[34]

Implementation of climate-smart policies, the World Bank further suggests, would safeguard development and help the region adapt to "whatever weather comes its way." The climate-smart policies promoted include a range of legal, infrastructural, and financial initiatives, envisioned and underway across the Western Balkans.[35]

A year later, the United Nations Development Programme (UNDP) revealed a climate change adaptation plan for the Balkans. Part of an "ambitious climate action across the world," the plan largely amounts to managing risks with climate finance.

Opening with dire forecasts—"annual flood losses in BiH are expected to increase 5-fold by 2050 and 17-fold by 2080"—the document lays out a series of plans to develop financing frameworks and solutions, including incentivizing the private sector to take part in climate action, for instance, by making insurance companies capable of accepting flood risks.

Optimistic projections of a bright future whatever weather comes are the norm for policy documents that discuss climate change with eyes on sustaining economic development. The Third National Communication, reporting on the country's greenhouse gas emissions and vulnerabilities to climate change, which the UNDP BiH prepared under the United Nations Framework Convention on Climate Change, gives a broader overview of the projected impacts of global warming on the country.

Included in the report are accounts of the high sensitivity of local ecosystems to the current and projected warming trends and expectations of endemic species extinction.[36] Nonetheless, the report reiterates the country's commitment to achieving sustainable development and envisions a "sustainable and prosperous 'green economy' by 2025, with low emissions, a high quality of life for everyone, preserved natural ecosystems, sustainable natural resources management and high level of climate resilience. . . . Negative impacts of climate change will be minimized by reducing vulnerability and taking advantage of opportunities brought about by climate change."[37]

These policy documents are genres of near-end ecologies. Extreme weather disasters figure prominently in forecasts that nonetheless remain bright because of a particular cosmology they presume. A shallow cosmology, focused on the plenitudinous present, does not look back in history to inquire about the past conditions of capital accumulation. The carbon economy that fuels development and recklessly drives the emissions of greenhouse gases in the meantime is entirely omitted from the account and along with it, questions of consumer, national, or corporate responsibility. Unburdened by history and focused on the immanent economic values at hand, these near-end ecologies can sustain the myth of development as the timeless and everlasting goal. Humans are presumed to be a lone species, outside ecological relationships and described obliquely by the metrics of the market: income and life standard. Vulnerability to disaster is likewise a property of a thoroughly economic being.

Issued as veritable prophecies, these end ecologies are not faithless, as they believe the capacity of capital to mend the ills, avert the threats, and eternally transcend the material and finite conditions of its generation. Nor are they godless. In the words of Ibn al-'Arabī: "There's no one who is not beseeching for something precious to them," nor is there anyone in dire circumstances who is not returning to God, however conceived, when all other possibilities are exhausted.[38] Brands of pure monotheism, the stories of weather disaster risk management devotedly invoke the one god of development by its many names: the green, the sustainable, the democratic.

The Prophet of Islam has bequeathed a prayer of protection to his community that says: I seek refuge in You from the accursed Shaitan, from his madness, his pride, and his poetry. The prayer suggests that Shaitan stalks even prophets; no one is immune to the devil's meddling. Poetry in the Qur'an is often a reference to well-composed, compelling messages that pose as Revelation. Shaitan's madness is his wish itself: instead of pleading for salvation or asking for forgiveness, while there is still time, he prays to bring others to ruin. He is in a rush, for he knows that every moment, the End is as close as ever. The root letters of the Arabic word for Shaitan, Zejd points out, also spell the word "neckbreaking." The Devil's earthly career earns him eternal fire, but the divine distance he already enjoys is the hell earned to him at present.

### *Apiculture through Drought*

"It's all about climate. Climate, climate, and climate. Climate!" Sead says, his bewilderment giving rise to this flustered rhyme. It is July 2017, and we are meeting at his apiary, situated on his paternal land in a village in Western Bosnia. For forty-five years, Sead has run a small apiary here with his wife's help. A small dacha by the hives' side makes the apiary the couple's favorite weekend retreat. Sead's knowledge of the local flora, its seasonal life cycles, and the ambient climate is deep, patiently built over years of observation. He says he remembers a different apicultural rhythm.

Throughout the mid-1980s, he and his wife would harvest the first batch of honey by the tenth of July—a mix of linden and meadow—then return the empty combs to the bees and drive to the Adriatic coast with their children for a family vacation. Back from the seaside, Sead would harvest another thirty kilograms of honey per hive before the end of August while also leaving the bees with plentiful honey for winter supplies.

That Sead's records of honey harvests are so closely matched to the memories of his family's vacationing schedule speaks to the ease with which beekeeping was once done. He remembers it as a sweet summer routine. "That was beekeeping, that was nature, that was paradise for both us and the bees! But now, a bee lives on sugar. What sustenance can it draw from sugar? Bare carbs, enough to move its wings and survive. But proteins, minerals, vitamins, everything vital is missing. So, you can add things up."

At a time when public discourses on global warming were rarely conducted in Bosnia, Bosnian Muslim beekeepers regularly reflected on the deranged weather in the language of climate.[39] The implications apiarists complained about were practical and immediately discernible. At the same time, because apiculturists' foraging year is contingent on the quality of the past summer's forage, the bees' wintering success, and the fact that each foraging season determines the outlooks of the year to come, the apiarists' perspective always extends to the near past and projects ahead, to the near future.

At the turn of 2017, beekeepers across the country reported exceptionally high winter losses, which most attributed to infections by the varroa mite, which were exacerbated by the poor quality of winter food supplies. Honey had been getting progressively scarcer since 2014. In 2016, beekeepers reported honey harvests to be 20 percent of their annual average.

Then came 2017, twice as bad. The signs of spring rushed in too early, prematurely initiating the plants' and bees' development. As the fruit trees blossomed, a heavy snowstorm blanketed the region in late April, devastating forage and trapping the populous bee societies inside their nests to feed on precious food stores.

In June 2017, the clement weather seemed promising, but then the heat rose, ushering in a drought. Temperatures spiked above 38°C (100°F), high enough to discourage Carniolan honeybee foragers from flying. Regional trees and plants such as linden and blackberry brambles are adapted to more moderate air temperatures. Their nectar fades above 32°C (89°F). The heat wave set in when the August meadow was expected to supply the bees' winter food stocks.

"Nothing smells, everything is so arid!" Sead says. He notes the loss of a sensuous clue that orients beekeepers to the local places and seasons, each marked by distinctive fragrances that trigger ambient moods and memories. Sead's remark is not sentimental but biographical and ecological. Floral scents are part of lived human local histories and make up apian memories, inherited and learned by foragers who use them to navigate to and assess the offerings of plants. The missing scent is an indication of dearth, and dearth's recurrence is a sign of a breaking bond between partner species.[40]

"So the bees are hungry and beekeepers will resort to feeding them sugar," Sead carries on.

In 2014, local beekeepers worried about feeding sugar to bees. By 2017, such emergency feeding had become routine. Artificial feeding manages the crisis, but, the local beekeepers worry, it compromises the bees' endurance in the long run and proves beekeeping unviable.

"All in all, our beekeeping is on a downward slope, I don't know how it will all end." As president of the local beekeepers' organization, Sead has a firsthand feel for the apicultural trends: "It expanded several years ago, young people were drawn into it, it was, well, promising, you could see a future in it, but now . . . You can't sustain the bees unless you turn to your savings, if you have any."

Contemporary biologists tend to describe honeybees as an extremely resilient species because of traits that helped this highly adaptive generalist pollinator spread and thrive across the earth's different climates. Compared with wild pollinators, *Apis mellifera*, the honeybee, is presumed to be better able to withstand the projected impacts of climate change, although the insect is already deeply compromised by intensive management and the anthropogenic environments. In fact, a rare attempt to speculate about climate futures of honeybees based on the assessment of the prevailing pressures upon the species state a concern that "climate-induced stress will in future compound the various factors already endangering the species in certain regions of the world."[41]

Such concern seem rather conservative especially considering the fact that there is a scarcity of experimental and field research on the subject, and that projections of pollinator futures rarely consider extreme weather events and extreme climate years.[42]

Local beekeepers' experiences give further reasons to doubt the confidence in honeybees' resilience to the warming climate. Weather extremes further aggravate the mounting effects of estranged seasons and weird weather on honeybees and their partner plants, and they thwart local beekeepers' best strategies for traveling and planting.

## *Human Responsibility*

"How has the year been, so far?" I ask.

"Worse than ever," Sulejman replies drily. My sister Azra and I meet this beekeeper in Western Bosnia in late July 2017. A thoughtful and versatile apiarist, Sulejman keeps his several hundred hives on the wheels for the length of the blooming season. Electronic hive scales message daily updates from the wide forage range, and he visits the field hives frequently to discern firsthand the current outlooks of the bees' forays. He also keeps hives closer to home, on the land that he enthusiastically plants with melliferous trees and plants, including phacelia, buckwheat, and lavender, and which he manages without synthetic pesticides.

His hives are currently set up in six locations, poised for chances of honey flow at different altitudes and across a range of microclimates, from riverine to mountainous. The flow, however, has been scant from the very beginning of the year.

"And now the drought! But the entire year has been catastrophic. It has brought the bees and beekeepers to the edge," Sulejman says, lighting up a cigarette. Azra rolls tobacco, and I see the two smokers tacitly bond over the silent pauses that smoking brings to a conversation. It is the first time we have met, and Sulejman, a thoughtful man in his late forties, gives us his sense of the times:

"The weather is becoming more extreme. Nectar and pollen losses over the year can now be total. The nectar has dried out at one thousand meters above the sea level. Plants there, too, are scorched! Imagine, then, what is it like [at the altitudes] below. Nectar is flowing somewhat in Trnopolje valley [in northwestern Bosnia] due to an abundance of small lakes and river channels. The plants look fresher there and the bees can collect at least enough for themselves. Otherwise, all's dry. It rained two nights ago, but the soil was so parched you could not tell the difference afterward. It didn't help the plants. It's an exceptionally hard year."

This apiarist tends not to speak much, so when he gets going, we do not interrupt.

Sulejman's beekeeping enterprise is exceptionally lucrative for the local standards because of the bee venom he has been professionally collecting for export to the North American pharmaceutical market.[43] Thanks to the venom export, Sulejman is not commercially dependent on honey harvests, nor is he pressed to finance the apiary losses with savings, regular incomes, or retirement checks, as is commonly the case with local beekeepers. This is the predicament that Sead was highlighting in the quote above. And yet a sound and steady supply of honey and pollen to the hives is indispensable for a sustainable venom collection venture, as only strong bee societies can afford to dispense and replenish a high expenditure of venom.

Sulejman manages his hives intensively, in the sense that he is doing his utmost to keep the bees' societies numerically strong, healthy, and fully employed throughout the nectar season. The care Sulejman shows for the bees' health and well-being, however, seems just as intense. Over the subsequent years of our acquaintance, I learned that he uses organic pesticides, makes his own supplemental food for the hives, and has made a number of adjustments to the design of standard bee boxes to give the bees a freer range in managing the nest's order and hygiene.[44] Given all his professional strivings, Sulejman's description of the year as "exceptionally hard" and of "the bees and the beekeepers on the edge" implies that apiculture is viable only insofar as the local ecologies allow for the seasonal syncing of weather, plants, and bees' cycles of growth, work, and rest.

Sulejman's apicultural practice also rests on a deeper sense of human responsibility and ecological connectedness between the human and the honeybee species.

"The fact that humans and bees are similar, is a sign (*išaret*) worth reflecting on. What is not good for the bees is not good for the humans, either. Because bees, too, have received the Revelation, just like we humans have. Just look how well their communities function. And we? A bee never lives just for herself but for the community and for the generations to come. And we? We think of the community narrowly—as of those gathered in a mosque for the Friday prayer. Community is whenever we gather to talk and act! We are wasting and polluting our nature, and someone ought to be able to live after us. How many believers (*mumini*) are aware of it? You know, for one, that our Prophet, peace be on him, advised us to use water sparingly. Whomever cares not to waste water while taking ritual ablutions is rewarded."

The point of departure for Sulejman's thinking is the presumed core similarity between humans and honeybees. The divine revelation that the two species have in common is the basis for thinking critically about human ways, but the species' affinity is also taken as an indication of their joined fates: struggling bees forebode trouble for humans, Sulejman suggests. His critique moves from humans in general to Muslims, and Bosnian Muslims, in particular: professing faith alone, Sulejman objects, is not enough to honor the gift of Revelation.

An Islamic community, according to Sulejman, is not confined to strictly ritual circumstances but is assembled on every occasion of shared words and deeds. The prophetic advice he cites is an example of how small ritual acts are steeped into responsible care for the earth and the elements that Islam inextricably connects to the rites of worship. Ritual ablution, being a prerequisite for an audience with God in daily prayers and for the handling of the Qur'an, Sulejman makes exemplary of the kind of wider ecological considerations that are expected of an ideal Muslim, the ideal embodied by the Prophet.

"*Iqrā*, we are told," Sulejman says, referring to what is generally taken to be the first command that God issued to the Prophet of Islam, "The incitement to learn, to take interest in knowing is not given for nothing. For nothing around us is random, there is no such thing as an accident. The traits of soil itself, say if it's acidic, determine which plants will grow and how well they will yield nectar. God is perfect and gives nothing but perfect arrangements in nature. And God is present on every square meter of the earth," Sulejman finishes his thought and follows it with a long pause. He smokes in silence.

The divine command, *iqrā*, is translated as "read" or "recite" and, more generally, "proclaim" or "transmit." The subsequent verses of the

Qur'anic chapter that begins with the command connect the recitation of divine revelation to God's generous teaching of the human *what he knew not*.⁴⁵ Qur'anic commentaries highlight knowledge as the quality that distinguishes *al-insān*, the human, from other species.

The wider meaning of *iqrā* is suggested by the translation of the Arabic original to BSC as *uči*, the imperative form of a verb that denotes learning, studying, and, among Bosnian Muslims, reciting the Qur'an. Sulejman's interpretation follows the lead in a wide sweep, which goes from the Book to nature, which in Islamic thinking figures as the material form of divine revelation, where sensuous things are signs of a divine active presence. Taking interest, for the *insān*, entails taking responsibility.

"The fact that we have placed the bees in hives makes us, makes me responsible and accountable to God. If the bees were over there in some tree trunk, it would be none of my business but since I have chosen their dwelling place for them, I am responsible to God for everything that the box implies. Because everything will testify against us or on our behalf. But how many people think that way? This is why it is often said that Sufis aren't living by the standards of this world, they have turned toward God, wholeheartedly. A few can understand them. Trouble is, so many people adore their own selves. They are at the mercy of their nafs. Shaitan has preoccupied them. They begin evaluating everything by means of money, run after profits."

The Qur'an says: *"You will receive what you wish for. If you desire the goods of this world, you will get them. If you desire the goods of the next world, you will get them." What really matters, at the end of the day, is to do some good to someone.*

Responsible apiculture is mindful of God's accounting. The perfect God keeps track of every meddling with the perfect order on Earth, in which He is present. *And God is enough as a witness.*⁴⁶ The *fasād*, the mess-making, Sulejman traces to one's self, the *nafs*. When a self reflects nothing but the self, occluding God's trace breathed into the *rūh*, the soul's bright counterpart, the human is an intimate with Shaitan. Intimacy whispers of desires for the goods and values of this world and the drives that wrack it. Implicitly, the drive to ruin has no other root than divine generosity, for whatever people want, the All-Merciful gives.

"You know what dervishes do in the Sufi lodges?" Sulejman asks next. "They are employed with grooming their souls and they do so by being at a service of a Shaykh. That's how it is with me and the bees: like a dervish, I am at their service."

Responsible beekeeping, effectively, opposes *fasād*. It resists Shaitan by working on the beekeeper's soul. Responsible beekeeping may not manage to save the bees, and it does not aspire to avert the world's collapse, but it presumes that the state of the world is inseparable from the affairs of the heart.

"We may be among the last generations of beekeepers," he says the last time we meet in 2020.

"You think?" I ask.

"We hope not, but it may just be so." He says it flatly, the composed man he is. Then he gets practical, advising me on how to build better bee boxes, the kind that are easy on the beekeeper's back and whose handling does not disrupt so much the bees' ways of dwelling.

## Notes

1. 50:16.
2. 30:41.
3. See, for instance, Ben Dibley, "Anthropocene: The Enigma of 'The Geomorphic Fold,'" in *Animals in the Anthropocene: Critical Perspectives on Non-Human Futures*, ed. Human Animal Research Network Editorial Collective (Sydney: Sydney University Press, 2015), 19–32; Clive Hamilton, *Requiem for a Species: Why We Resist the Truth about Climate Change* (New York: Earthscan, 2015); Foster, *After Sustainability*.
4. In some areas, 250 to 300 liters of water poured per square meter, per day. The rainfall was the heaviest recorded ever since meteorological stations in the country started tracking precipitation rates in 1894.
5. BH Pčelar, "Pogledi u nebo i molba Svevisnjem, Sulejman Alijagic," *BH Pčelar* 36, June 15–August 15, 2014.
6. Ibid.
7. Brood nest temperature is regulated in the range from 33 degrees Celsius to 36 degrees Celsius. The brood initiated in late winter is significantly smaller in size and kept by the clustering bees at 33 degrees Celsius. Development of the brood at suboptimal temperatures has a range of implications, including higher mortality inside cells, shortened longevity of adult workers, impairment of short-term learning and memory of adult workers, altered wing morphology, and disease prevalence. See Julia C. Jones et al., "The Effects of Rearing Temperature on Developmental Stability and Learning and Memory in the Honey Bee, *Apis mellifera*," *Journal of Comparative Physiology A* 191 (December 2005): 1121–1129, https://doi.org/10.1007/s00359-005-0035-z; Qing Wang et al., "Low-Temperature Stress during Capped Brood Stage Increases Pupal Mortality, Misorientation and Adult Mortality in Honey Bees," *PLoS* 11, no. 5 (May 2016): e0154547, https://doi.org/10.1371/journal.pone.0154547; K. Tan et al., "Effects of Brood Temperature on Honey Bee *Apis mellifera* Wing Morphology," *Acta Zoologica Sinica* 51, no. 4 (2005): 768–771.

8. BH Pčelar, "Priroda da se okrenula protiv nas ili mi protiv nje? Nista ne bi od 'zlatne godine,' Rajko Radivojac," *BH Pčelar* 36, June 15–August 15, 2014.

9. Islamic Community, *Islamska Zajednica*, is an institution that formally represents Muslims of BiH as well as Muslims in the wider region, including the countries of former Yugoslavia and Hungary and the diasporic Bosnian Muslims. For a helpful summary on the Islamic Community's organization and history, see David Henig, *Remaking Bosnian Muslim Lives: Everyday Islam in Postwar Bosnia and Herzegovina* (Champaign: University of Illinois Press, 2020).

10. 114:5.

11. 2:3.

12. 41:46.

13. 21:47.

14. For a thoughtful discussion of *khilafah*, human vicegerency on Earth, see Tlili, *Animals in the Qur'an*, 115–123. For comparison, a more conventional reading is found in Richard Foltz, "'This She-Camel of God Is a Sign to You': Dimensions of Animals in Islamic Tradition and Muslim Culture," in *A Communion of Subjects: Animals in Religion, Science, and Ethics*, ed. Paul Waldau and Kimberly Patton (New York: Columbia University Press, 2009), 149–159.

15. BiH is apparently among the few countries left in Europe where phosphate concentrations in detergents are high (up to 30 percent) and unregulated. See Center for Ecology and Energy, "Deterdženti bez fosfata—napredak za okoliš," Projects, posted January 2012, https://ekologija.ba/2017/05/22/deterdzenti-bez-fosfata/.

16. Ibn al-'Arabi, *Mekanska Otkrovenja*, 62–74.

17. Industrial and household waste management in the country is nothing short of catastrophic. Unregulated, so-called wild (*divlje*) landfills are found in forests and in uncultivated meadows across the country. Town landfills are maintained in environmentally unsound locations and are entirely unequipped to handle the industrial and medical waste that they regularly receive. In addition, industrial enterprises deposit waste on their compounds or by the riverbanks or incinerate it in open pits, their fires visible to the passersby from the highway. Defunct industrial zones across the country store extremely toxic waste in shallow mounds or in containers that have deteriorated over time. Moreover, *kruks*, the by-product of toluene diisocyanate (TDI), used in the former chlor-alkaline industrial complex, has reportedly been confused for gravel and used for local road construction. See, for instance, Armin Kendić, "Kruks, živa i hlor u industrijskoj zoni: Opasan otpad prijeti zdravlju gradjana Tuzle," *Klix*, March 5, 2016, https://www.klix.ba/vijesti/bih/kruks-ziva-i-hlor-u-industrijskoj-zoni-opasan-otpad-prijeti-zdravlju-gradjana-tuzle/160304151.

18. In addition to the obligatory daily prayers, there are special-occasion and recommended forms of salat, passed down by the Prophet, which may be performed to seek divine closeness and pleasure or in special circumstances when seeking guidance, in the case of an ardent wish or a need for protection.

19. 51:49.

20. 24:21; also 6:42.

21. 26:89.

22. 91:9–10. See the brief commentary in Seyyed Hossein Nasr et al., eds., *The Study Qur'an: A New Translation and Commentary* (New York: HarperCollins, 2017), 1520.

23. 59:16.

24. 14:22.

25. 2:11.

26. During 2017's summer, peak hospital admissions were reported across Europe. Especially vulnerable are the young, the elderly, outdoor workers, and those with no access to cooling devices or facilities—which includes the majority of the population in BiH. The effects of extreme temperatures are aggravated in built environments where pavement and

structures can develop heat from 10°C to 37°C (50°F–100°F) higher than the air while urban "canyons" add anywhere from 3.9°C to 6.6°C (7°F–12°F) to the city heat load. Air quality is also affected since hot and sunny days tend to increase ozone levels and, in turn, to spike levels of NOx gases, which cause damage to human respiratory health and ecosystems through the formation of smog, ground-level ozone, and acid rain. See Jay Lemery and Paul Auerbach, *The Enviromedics: The Impact of Climate Change on Human Health* (Lanham, MD: Rowman & Littlefield, 2017), 23. Furthermore, persistent heat combined with drought tends to increase energy use: electricity consumption on the Adriatic coast broke all records in August when Lucifer checked in as scores of overheated inland Croats traveled to the seaside to cool off. H., "Lucifer poharao Europu: najmanje dvoje ljudi umrlo od vrućine, najteže pogođeni Balkan i Italija, na Jadranu i danas paklenih 42 stupnja," *Slobodna Dalmacija*, August 4, 2017, https:// slobodnadalmacija.hr/vijesti/svijet/lucifer-poharao-europu-najmanje-dvoje-ljudi-umrlo-od -vrucine-najteze-pogodeni-balkan-i-italija-na-jadranu-i-danas-paklenih-42-stupnja-500315. Extreme heat also overburdens power lines, compromises power plants that use water coolers for safe operations, and disables or hampers hydroelectric plants—Albania's hydropower, for instance, had to shut down, forcing the country to import 80 percent of its electricity needs. Center for Climate and Energy Solutions, "Heat Waves and Climate Change," Climate Basics, Extreme Weather, accessed October 17, 2019, https://www.c2es.org/content/heat-waves-and -climate-change/; Elizamar Ciríaco da Silva et al., "Drought and Its Consequences to Plants— From Individual to Ecosystem," in *Responses of Organisms to Water Stress*, ed. Sener Akinci (London: IntechOpen, 2013), 17–47.

27. Instead, various forms of assistance and insurance payoffs were organized sporadically on the level of cantons, while the Serb Republic committed €10 million toward fall sowing subsidies and irrigation systems. Given that 50 percent of the agricultural output in BiH, estimated at a value of €300,000 million, was lost, the aid was far from enough but still appreciated. "It's a shame that similar measures were not undertaken in the Federation" said the minister of agriculture of the Serb Republic. Udruženje poljoprivrednika u Zeničkodobojskom Kantonu, "Apel Vladi Federacije za hitno proglašavanje stanja elementarne nepogode," *Zenicablog*, August 5, 2017, https://www.zenicablog.com/apel-vladi -federacije-za-hitno-proglasavanje-stanja-elementarne-nepogode/; Rahela Čabro, "Teško ljeto za poljoprivrednike u BiH, suša uništila veliki broj usjeva," TNT portal, August 11, 2017, https://tntportal.ba/vijesti/tesko-ljeto-za-poljoprivrednike-u-bih-susa-unistila-veliki-broj -usjeva/; BN televizija, "Nema elementarne nepogode?!," *BN televizija*, August 11, 2017, https:// www.rtvbn.com/3875127/nema-elementarne-nepogode. Consequences were dire for cattle farmers as well. Dairy cows produced less milk, and, worse, farm animals refused food. The heat caused cattle miscarriages, poor conception rates, and deaths; its longer-term effects are yet unknown.

28. Disastrous flooding events on the continent were recorded in 2002, 2005, 2010, and 2013. The effects of extreme precipitation are often exacerbated by flood waves or snowmelt, and regularly so by the country's mountainous terrain and porous rock bed, which are highly susceptible to feeding flash floods. Sava River management requires cross-border coordination among the three ex-Yugoslav states on its banks, whose political relationships remain uneasy. Its management was further hindered on the Bosnian side as the rivershed infrastructure was damaged in the war and has further deteriorated since the peace due to neglect. Another assessment of vulnerabilities that followed in 2011 was meant to provide a blueprint for the national disaster protection and response plan. Conducted by a team of experts drawn from both entities of the bipartite Bosnian state (the Croat-Muslim Federation and the Serb Republic), the assessment also highlighted flood risks. Its conclusion reads: "Taking into consideration the string of extreme weather events over the last ten years as well as the future projections of climate change, we can conclude that floods present the greatest danger to the community." Ministry of Security of Bosnia and Herzegovina, "Procjena ugroženosti Bosne i Hercegovine

od prirodnih ili drugih nesreća, 2011," Documents, Other Documents, posted March 31, 2014, http://www.msb.gov.ba/Zakoni/dokumenti/default.aspx?id=10773&langTag=bs-BA. Two years later, an update on flood prevention activities in the Federation found evidence of systemic disinvestment from the essential basin and river management infrastructure, diversion of flood prevention funds to the government's budget, incompetence and negligence in the highly fragmented structure of watershed governance, and administrative and political hurdles to coordination within and between the two entities of the Bosnian state. Audit Office for the Institutions of the Federation BiH, "Izvještaj revizije učinka prevencija poplava u Federaciji BiH, 2013," Performance Audit, posted January 21, 2013, https://www.vrifbih.ba/?s=Izvještaj +revizije+učinka+prevencija+poplava+u+Federaciji+BiH,+2013&post_type=.

29. The economic impact of the 2014 May storm was grave: the losses and damages, estimated at €2.04 billion, made up nearly 15 percent of the national GDP. The brunt of the losses (75 percent) was suffered by the private sector. The majority of destroyed homes, businesses, and farms and crop fields were not ensured, which means that the losses the owners suffered were unrecoverable. The "Recovery Needs Assessment" acknowledged a number of "human factors" at fault in aggravating the floods and precipitating landslides, including "uncontrolled exploitation of forests and minerals, [and] an increase in illegal and or unplanned construction." ILO (International Labor Organization), "Bosnia and Herzegovina Floods 2014: Recovery Needs Assessment," Report, p. 23, posted June 30, 2014, https://www.ilo.org/global/topics /employment-promotion/recovery-and-reconstruction/WCMS_397687/lang--en/index.htm. At the same time, the document was silent on the more formally political preconditions for the disaster, such as the failure of the state to maintain the watershed infrastructure, coordinate flood prevention strategies across the entity and state borders, or implement monitoring and early warning systems, along the lines that the audit report had already recommended in 2012. Put simply, anthropogenic factors were entirely depoliticized and privatized: the blame was implicitly laid on irresponsible anonymous actors while the state's roles in watershed management and regulating construction and natural resource extraction were left unimplicated. Ibid., 64, 253.

30. Ibid., 23.

31. Ibid., 6.

32. Van Gelder, "It Is Time for Action on Climate Risk in the Balkans," The World Bank, Opinion, posted on September 17, 2018, https://www.worldbank.org/en/news/opinion/2018 /09/17/it-is-time-for-action-on-climate-risk-in-the-balkans.

33. Ibid.

34. Ibid.

35. Serbia is singled out as "blazing" a path of the climate-smart development with its new legislation framed around disaster risk management and international borrowing to minimize the impacts of catastrophes and speed up recovery. Serbia had rehauled its legislation around the disaster management framework (in alignment with Sendai Framework) and secured a sixty-six-million-euro loan from the World Bank to ensure access to the recovery fund. The loan's special feature is "Catastrophe Deferred Drawdown Option," an instrument "designed to protect vital social investments from the fiscal impacts of natural disasters." The World Bank is not the only institution impressed by these achievements. Citing Serbia as a model, an opposition party member in the House of Representatives, in the Parliament of the federation of BiH, in cooperation with civil sector and international agencies drafted a bill on risk reduction and disaster management. See ibid.

36. UNDP Bosnia and Herzegovina, "Third National Communication (TNC) and Second Biennial Update Report on Greenhouse Gas," Publications, posted July 12, 2017, 105–107, https://www.ba.undp.org/content/bosnia_and_herzegovina/en/home/library/environment _energy/tre_i-nacionalni-izvjetaj-bih.html. Another UNDP assessment suggests that "average temperature increase greater than 2°C (36°F) will result in costly adaptation, and impacts

that will exceed the adaptive capacity of many ecological systems (such as high mountain and lowland oak forest areas), and a high risk of large-scale irreversible effects including endemic species extinction." UNDP Bosnia and Herzegovina, "Climate Change Adaptation and Low Emission Development Strategy for BiH, 2013," Research & Publications, Energy and Environment, January 9, 2014, 20, https://www.ba.undp.org/content/bosnia_and_herzegovina/en/home/library/environment_energy/climate-change-adaptation-and-low-emission-development-strategy-.html.

37. UNDP Bosnia and Herzegovina, "Third National Communication," 162.

38. Ibid., 459–460.

39. One obvious change in local news reporting on weather, since the May 2014 floods, has been the integration of extreme weather monitoring and early warning systems. In keeping with typical sign systems elsewhere, color-coding became the standard shorthand for extreme weather emergencies and implicitly stood for preparedness and resilience, but the broader contexts of global change that yellows, umbers, and reds were signaling were left out of the reportage's code. With the exception of a few published conversations with beekeepers, climate change at the planetary level made no local or regional news during the drought. FENA, "SUŠA DRASTIČNO SMANJILA PROIZVODNJU MEDA: Tržište će da preplavi VEŠTAČKI," *Blic*, August 16, 2017, https://www.blic.rs/vesti/republika-srpska/susa-drasticno-smanjila-proizvodnju-meda-trziste-ce-da-preplavi-vestacki/wjqmf6k; Al Jazeera Balkans, "Teška godina za pčelare u regionu," *Al Jazeera Balkans*, July 30, 2016, http://balkans.aljazeera.net/vijesti/teska-godina-za-pcelare-u-regionu.

40. Relatively little is known about the effects of climate warming on plant physiology and the production of floral smell and nectar. Experimental and field studies showed a reduction of sugar content and nectar volume in the heat-stressed plants as well as a production of flowers without nectar (Theodora Petanidou and Erik Smets, "Does Temperature Stress Induce Nectar Secretion in Mediterranean Plants?," *New Phytologist* 133, no. 3 [July 1996]: 513–518, https://doi.org/10.1111/j.1469-8137.1996.tb01919.x). Experiments involving Mediterranean plants found results varying depending on the species and on water availability (Krista Takkis et al., "Climate Change Reduces Nectar Secretion in Two Common Mediterranean Plants," *AoB Plants* 7 [September 2015]: plv111, https://doi.org/10.1093/aobpla/plv111). A team of Europe-based researchers is specifically investigating floral scent alterations in heat-stressed plants and the ensuing changes in plant-pollinator interactions. See Coline Jaworski, Benoît Geslin, and Catherine Fernandez, "Climate Change: Bees Are Disoriented by Flowers' Changing Scents," *The Conversation*, June 25, 2019, https://phys.org/news/2019-06-climate-bees-disorientated-byflowers-scents.html. Petanidou and Smets, "Does Temperature Stress Induce Nectar Secretion in Mediterranean Plants?" 513–518; Ettore Pacini and Massimo Nepi, "Nectar Production and Presentation," in *Nectaries and Nectar*, ed. Susan W. Nicolson, Massimo Nepi, and Ettore Pacini (Dordrecht: Springer, 2007), 167–214; Victoria Scaven and Nicole E. Rafferty, "Physiological Effects of Climate Warming on Flowering Plants and Insect Pollinators and Potential Consequences for Their Interactions," *Current Zoology* 59, no. 3 (September 2013): 418–426.

41. Yves Le Conte and Maria Navajas, "Climate Change: Impact on Honey Bee Populations and Diseases," *Revue Scientifique et Technique* 27, no. 7 (August 2008): 507, http://dx.doi.org/10.20506/rst.27.2.1819.

42. See Pablo Marquet, Janeth Lessmann, and Rebecca Shaw, "Protected-Area Management and Climate Change," in *Biodiversity and Climate Change: Transforming the Biosphere*, ed. Thomas E. Lovejoy, Edward O. Wilson, and Lee Hannah (New Haven, CT: Yale University Press, 2019), 283–293. For an overview of issues on alterations in plant-pollinator phenologies in relation to climate change, see Joar Stein Hegland et al., "How Does Climate Warming Affect Plant-Pollinator Interactions?," *Ecology Letters* 12, no. 2 (February 2009): 184–195, https://doi.org/10.1111/j.1461-0248.2008.01269.x; Camille Parmesan, "Influences of Species, Latitudes and Methodologies on Estimates of Phenological Response to Global Warming," *Global*

*Change Biology* 13 (August 2007): 1860–1872, https://doi.org/10.1111/j.1365-2486.2007.01404.x; and Eric Post and Michael Avery, "Phenological Dynamics in Pollinator-Plant Associations Related to Climate Change," in *Biodiversity and Climate Change: Transforming the Biosphere*, ed. Thomas E. Lovejoy and Lee Hannah (New Haven, CT: Yale University Press, 2019), 42–55. A comprehensive overview of honeybees' resilience and vulnerabilities and speculations on apian climate futures is offered by Yves Le Conte and Maria Navajas, "Climate Change: Impact on Honey Bee Populations and Diseases," *Revue Scientifique et technique* 27, no. 7 (August 2008): 499–510, http://dx.doi.org/10.20506/rst.27.2.1819.

    43. Bee venom is an expensive substance and is of special interest to the skin-care industry for its stimulative effect on collagen production. Its primary therapeutic use is for the treatment and management of arthritis, neuralgia, multiple sclerosis, and lupus. Apitoxin's active components and pharmacological properties have been researched around the world since the 1950s. Composed of proteins, enzymes, polypeptides, minerals (sulfur among them), and neurotransmitters, including dopamine and serotonin, bee venom has been associated with anti-inflammatory effects. These effects are especially attributed to melittin, a polypeptide agent composed of twenty-six amino acids said to be far more potent than hydrocortisol. See Jack Gauldie et al., "The Peptide Components of Bee Venom," *European Journal of Biochemistry* 61, no. 2 (January 1976): 369–376, https://doi.org/10.1111/j.1432-1033.1976.tb10030.x; Jun Chen et al., "Melittin, the Major Pain-Producing Substance of Bee Venom," *Neuroscience Bulletin* 32 (March 2016): 265–272, https://doi.org/10.1007/s12264-016-0024-y; Obioma Eze, Okwesili Nwodo, and Victor Ogugua, "Therapeutic Effect of Honey Bee Venom," *Journal of Pharmaceutical, Chemical, and Biological Sciences* 4, no. 1 (March–May 2016): 48–53, https://www.semanticscholar.org/paper/Therapeutic-Effect-of-Honey-Bee-Venom-Eze-Nwodo/4be85dc94738742f9102b2694273835d57a8ceb2; Abdul Rahim Al-Samie and Mohamed Ali, "Studies on Bee Venom and Its Medical Uses," *International Journal of Advancements in Research & Technologie* 1, no. 2 (July 2012), https://pdfs.semanticscholar.org/a0b1/f2ed65fc1b999b657045135d563e7bfd298b.pdf?_ga=2.55221038.485531400.1581172971-2113641594.1577775850; Jojriš, *Pčele i Medicina*, 127–129.

    44. He breeds his own honeybee queens to restock the hives every year and keeps societies employed by producing the brood to fortify or restock his foraging hives. Sulejman mistrusts commercial pollen substitutes and makes his own from the supplies of pollen he gathers from his hives throughout the spring.

    45. 96:6.

    46. 48:28.

# − 5 −

# *The End*

## God's Promise

When the time comes, Isrāfīl blows the horn. An attentive angel, tradition has it, his luminous body is covered with down. I imagine bristles finer than the bees' and just as receptive. The horn has long been poised to his countless mouths, poised all across his limbs. Wakeful, Isrāfīl has not once blinked for fear of missing the sounding cue. An immense angel, but his compassion is far greater than his size. Every day and once a night, he is overwhelmed by sorrow at the sight of hellfire, and he rains tears. God, the Compassionate, has appointed such a tender creature as the herald of the End. The end that, for all we know, may be near. The Qur'an cites God as saying, *Truly, the Hour is coming, no doubt about it.*[1]

Both death and the end of the world are God's promises. *And the death throes will come in truth; this is what you have been trying to avoid. And the Horn will be blown. This is the promised day.*[2] The Book repeatedly brings them together, and Islamic textual tradition follows suit by insistently joining the two themes. God calls it simply *the day, the hour, the event,* or *the term*, as well as, more imposingly, *the Catastrophe (al-qāri'ah)*. Often, *al-qiyāmah* (the standing)—while awaiting divine judgment—stands for all other terms and is transliterated as *kijamet* (pronounced *keeyamet*) in the Bosnian Muslim dialect of Bosnian-Serbian-Croatian (BSC).

*Apocalypse*, a term loaded with Judeo-Christian history, which denotes unveiling and suggests the utter moment when the world is torn apart and the subject is fully present to the visiting God, can only serve as a loose translation. The freight of discomfort and fascination that *apocalypse* carries makes it an indispensable word, irresistible not least due to its present invocations, reinterpretations, and denouncements in the context of global environmental and climate catastrophe. *Apocalypse*, however, often strikes ears too flatly, its meaning drained by its prolific past uses, the warning it invokes emptied. I use the word cautiously, therefore, without wanting to disown it, especially not its Abrahamic lineage or its capacity to prophesize or threaten the possibility of the world's doom.

The point of this final chapter is to bring to the fore the Islamic sense of the End, in the language of the Qur'an and the idiom of Bosnian Muslims, with the hope that the subject of finitude and the end-time prophecy—the old news and perennial themes—can be heard somewhat anew now that our biosphere is being fast undone.

To begin with, I attend to the Qur'anic conflation of personal death and the world's devastation; their pairing makes the strongest impression on listeners. I then focus on the fact that both are divinely promised, and *surely, God does not break His promise*.[3] The possibility of the End looms large, inspires dread as well as hopeful expectations, and, just as often, incites calm contemplation of one's own death. Remembrance of death affords self-awareness, moments of return to God, cultivated or gained and lost, on the go.

Understanding the role of eschatology as both a mundane and profound reference for living a Muslim life helps make sense of the special significance that honeybees carry for the locals as beings vital to the world. The loss of their humming, honey-making, and dwelling—not just of their pollination service—is an omen of the growing unsustainability of the world. In statements that are taken to be self-evident and in propositions that are unfinished, suggestive fragments, apian and human fates are rendered inseparable.

The previous chapters showed how beekeepers, through practice, tune into insect distress audible in the wheeze of withered plants and discernible in diminished nectar, in the loss of flower fragrance, in the drawl of unseasonal elements. This chapter shifts attention away from the empirical arguments that people make for ecological disasters due to climate change. Instead, the chapter follows local contemplations, some of which are left unfinished, conceptually rough, or merely suggestive, and carries them onward through readings of and conversations about the Qur'an.

The Qur'an has inspired expansive meditations on matters of death, the end of this worldly life, and the eternal life to follow. Such meditations are expansive because they presume and elaborate on both Islamic cosmology, in which humans are one special kind among many beings, and Islamic metaphysics, in which the fabric and meaning of humans and other beings can only be grasped by attending to divine acts and attributes in relation to the cosmos. This breadth and depth make eschatology the most relevant theme in Bosnian Muslim stories.

Stories are often treated as useful tools because they help teach, describe, inspire, or better summon attention to some subject of concern, but they are presumed to be entirely fictional or symbolic, and, therefore, unbinding and open ended. Storytelling as such has become very popular in the contemporary arts, humanities, and social science scholarship as a means of retelling the material reality in deliberately enchanting or inventive ways, to expand modern notions of what the world is, and to critique or counter the destructive economic, technological, and interspecies relationships that rest on this confidence. Wonderful accounts of stones and weeds, elements and plastics, mushrooms, ghosts, and gods do expand the horizon of social theory and inspire readers to look at and relate to the world differently and imagine a hopeful possibility of an entirely different politics and planetary future.[4] These inspiring stories, however, frequently betray discomfort or impatience with traditional, "old," monotheistic religious myths, especially if such myths convey certitude about the world's apocalyptic end.

On the other hand, the expansive cosmological and metaphysical elements of Qur'anic and prophetic tradition seem to embarrass those Muslims who trust that rationalist, naturalistic, scientific—in short, modern—narratives are the most economical, sensible, and effective ways to render the religious tradition relevant to contemporary Muslim lives and comprehensible to the secular or non-Muslim audience.

To the storytellers of both strides, the inventive and the sober, and to anyone else who cares to listen, I offer an ethnographic account—relating what people have said, read, thought, and done—with story elements that are mythical but indispensable and anything but arbitrary because they draw on and elucidate the Revelation. The book started with the tale of two angels who worryingly inquired about the state of this world. Those two angels suspected that the decline of honeybees, in particular, indicated the near end.

This chapter opens with Isrāfīl, one of the four great angels in Islamic cosmology, who is also awaiting the End. But this angel's greatness also has to do with his foresight, as Isrāfīl is not worried about this world. With ears pricked for God's forthcoming cue, his eyes drift ahead to the next world: *al-ākhirah*.

With the awesome horn at the ready, angelic alertness, gentleness, and grief are modes of listening to God's speech that restate the keyword of human responsibility in the light of finitude and eternity, the constant themes of Revelation. When the horn sounds—and the time is always ripe—what ends is the world that has failed to sustain, through listening, a divine desire that keeps up the world and flows with honey.

## *A Small and the Great Kijamet*

"When a person dies, kijamet already sets in for him," Shaykh Ayne says, retelling a saying attributed to the Prophet. I had just finished reading aloud from al-Ghazālī's book on the remembrance of death over our meal.[5] We often dined over texts in the kitchenette adjacent to Shaykh's home library.

"Love whomever you will," al-Ghazālī cites the Prophet, "but know that you'll leave him."[6] Shattering prose for anyone in love. Remembering the eventuality of death is meant to intensify and improve the quality of one's relationships in and to the present, lest we quarrel or take each other for granted.

But remembering death is also meant to spoil one's pleasure, the Prophet has said. For everything in this life will come to pass, and the ephemeral, as sweet as it may be, should not detract attention from the eternal life to come. In this sense, the Prophet calls death a "precious gift to a believer" as well as a sufficient coach through life.[7]

One's own death appears most improbable, al-Ghazālī writes, or at least seems to be far off in the future until, one day, it surely comes as a surprise. So, while there is still time, al-Ghazālī advises, let the human think "about his body and limbs that, inevitably, will be eaten by worms, that bones will rot. Let him think which of his lenses will the worm eat first: the right one or the left."[8] Decaying body and passing time, al-Ghazālī suggests, are part of the near world, *dunyā*, while "knowledge and sincerely good deeds" remain forever because their record shapes one's course through the next world, *ākhira*.[9]

The Qur'an repeatedly reminds its readers and listeners of their coming death and pairs up the imagery and evoked experience of dying with the event of *qiyāmah*. A verse frequently cited reads: *Every soul will taste death and you will only be given your full reward on the Day of Qiyāmah. So, whomever is drawn away from the Fire and admitted to the Garden has succeeded. And what is the life of this world except the enjoyment of delusion.*[10]

The taste of death is the last thing a soul tastes in this life. Worldly life, no doubt, offers tastes of pleasure, and the Qur'an encourages Muslims to enjoy rather than deny themselves the worldly joys—tasty food, perfumes, fine clothes, sexual play—which are God's gifts for as long as they indulge in the permissible and with a good measure. One's last taste, however, proves all pleasures deceiving insofar as they have came to pass and their very impermanence is a sign of their partial reality, their essential nonexistence, and their contingency on the everlasting God, the Living and the Self-Subsisting. By bringing together the events of personal death and *al-qiyāmah*, this verse, well loved among Muslims, reminds the listeners that the tasting—which is the immediate, visceral way we grasp pleasure and pain—is a mode of experience that carries on, with the soul, after death and that the eternal quality of this experience is determined on the day of rising and judgment. Death, in short, proves finitude and fulfills a divine promise to "every soul."

Shaykh Ayne's comment, however, suggests an even more vital relationship between death and the world's end. Death is already the world's end, he says, for the person dying, and the world's end is somehow like an individual death. The two are sometimes described as the small and the great term, event, hour, day, or, indeed, *al-qiyāmah*.

The Qur'anic chapter titled "Al-Qiyāmah" helps me follow Shaykh's thread. It also links death and the End but does so in reverse order from the verse above, from the torn skies to the worn-out body. It begins with God swearing by the Day of *Qiyāmah* in response to the people's mocking disbelief in its coming, then depicts the day when Earth is shattered: *When the sight is struck, and the moon eclipsed, and the sun and the moon joined—on that Day, a human will cry: "Where is there to flee?" No way! There is no refuge. With Your Lord Alone is the comfort, that day.*[11]

From the day of Earth's doom, the chapter fast-forwards to the Day of Judgment, when the record of everyone's past deeds is reviewed: *Human shall be informed that day of that which he has sent forth and of that which he has left behind.* The next verse takes us to a deathbed: *When the soul has reached the collar bones and it is said "Who can cure him?" and knows it is the time to part, and a leg twines with a leg, to your Lord that day it will be driven.*[12]

The act of dying existentially, unwillingly, enacts *shahādah*, the quintessential Islamic truth claim—*lā ilāha illā Allāh*—there is no god but God. I am no god, after all, since I am dying. Isrāfīl's horn blasts much the same message to the whole cosmos when, finally, all its beings, including the angels in the near heavens, perish, testifying in the act that they are likewise no gods: have no life, no power, no existence of their own. Finitude is always on the stage of this world, so even a distracted audience takes notice sooner or later. But the human self, more than anything else, acts and imagines itself godlike, unwilling or unable to conceive of its life as anything but enduring. Death delivers a divine promise to each and all. At the same time, the Qur'an suggests, when one breathes one's last breaths, the agony comes with revelations: that the soul is about to transcend the body—*it is time to part*—and that there is nowhere to run but to "Your Lord." *'Illa allāh*, but God.

Personal death is the strongest argument the Qur'an makes for the inevitability of the world's end, the world being likewise a subject in relation to God whose end is sworn. So when the oceans boil over; when the skies split, red like roses and as murky as burned oil; when the stars dim, the whole, vast universe is folded into nonexistence, *like writing scrolls are rolled up.*[13] When the noble and close angels perish and Isrāfīl's splendid horn is put to rest at his hirsute side, then, finally, the angel of death ʿAzrāʾīl is asked to extinguish himself. "Dear Lord, help me out," he pleads, raising his hands. Having dealt death on others since the dawn of time, the angel finds himself all thumbs when it comes to his own "I." When all is rolled back from manifest reality into the realm of possibility within divine knowledge, God asks, *To whom belongs the power today?* None is left to reply. *illā Allāh*, except God: *To God, the One, Prevailing.*[14]

The Qur'an conveys the voice of God: *We created not the heavens and the earth, and that between them, in play. We did not create them except in truth; but most of them do not know it.*[15] The formal Revelation, to be taken on faith or doubted, expresses the truth and reality of God, the Real or the True, being among God's names (al-Ḥaqq), and the meaning of life in the light of that truth and the eternal life. The Day of *Qiyāmah*, make the truth manifest, the Qur'an claims, not just out in the open but also, profoundly, within: *On the Day when the [horn] blasts, and the second blast follows, hearts that day shall tremble.*[16]

Because time is fleeting and death is a stalker, Al-Ghazālī recommends that his reader contemplate death with a "present heart." Shaykh Ayne explains how: with *zikrullah daimen*, he says, keeping up an ongoing (*daimen*) remembrance of God, the source of presence.[17]

## *Listening*

Al-Ghazālī, the famous medieval Islamic scholar and Sufi, remains Shaykh Ayne's favorite source ever since he read him in Arabic as a young man of fifteen, but reading this book aloud over dinner was a mistake I soon regretted. As a long-time student and teacher, I tend to treat texts as readable anywhere, at any time. To treat books like teaching and learning tools presumes a great deal of appreciation for texts but also implies a callousness of sorts, a thick-skinned sort of reading due to casual handling of citational sources.

Shaykh Ayne started seriously associating with books at a younger age than I did. At five, he was reading the Qur'an in Arabic in the warm glow of candlelight at the turn of the 1940s, under the guidance of his father, an imam and a beekeeper, the way his father had learned from his father, and so on, throughout the generations of this imamic family, which traces its roots to an ancestor appointed to Western Bosnia by the Ottoman Sultan Mehmet II in the fifteenth century.[18]

Once the boy's Qur'anic lessons were completed, his range of readings in Turkish and Arabic expanded to include core texts in Islamic jurisprudence and creed as well as great Islamic exegetes and thinkers, including al-Ghazālī. In 1970, Ayne was assigned to the post of imam in a small semi-urban mining community and secretly embarked on the Sufi path, becoming a dervish. Secretly not only because the socialist government was intolerant of Sufi lodges and practices but also because some Sufis shun public displays of devotion.

His appointment left him time to read, but his day-to-day job also presumed he would tell stories to the members of his mosque community, the lore that draws on the Qur'an but weaves its revelations into a rich kilim of characters and narratives from cosmology and sacred history, from the Prophet's life and his community, or from biographies and legends about memorable Muslims through the ages.

As a dervish, Ayne began training with Sufi masters, which, first of all, presumes listening. His first Shaykh was a carpenter by day with a basic knowledge of Quranic Arabic and neither the time nor interest in reading anything other than the Qur'an. Ayne's two other guides were close friends and giants of twentieth-century Bosnian Sufi thought, both extremely well read. One was a brilliant mathematician who cherished anonymity, and the other, Shaykh Mustafa Čolić, was an imposing figure and an exuberant character, an imam and speaker with a public profile, not the kind to hide. Their citations ranged from the classics of Islamic theology and philosophy to their contemporary interpreters and the canon of European philosophy.

For these two intellectuals, the Qur'an was the primer and the cornerstone for their writings, reflections, and teachings. A dervish listens to the guides who know more than he does—dervishes in Bosnia are rarely women—and the listening, which is a demanding task, is a way of training. A genuinely good storyteller is one who, first of all, has learned how to listen.

The stories heard and told are returning to and reopening the Qur'an, the vibrant reference, with its 114 chapters and whose literary quality in the original Arabic is both stunning and abundant in interpretive possibilities, as the Book's countless commentaries in prose and poetry testify. The Qur'an as the fount of storytelling is binding, but, at the same time, the tradition is capacious, giving rise to always new—and newly indebted—accounts and inflections.

Shaykh Ayne's Sufi teachers in particular were expounding on the potential for a multitude of retellings and hearings, the potential that draws on the nature of reality itself as an ample and growing plenitude, *kathrah*, the domain of ongoing divine self-revelations. The unique God discloses His uniqueness by engendering singularities: things, events, beings, and words that are irrepeatibly distinct. Divine self-disclosure yields prolific manyness from without and within: *We will show them Our signs in the horizons and within themselves*[19] so that the listeners themselves are not only different from one another but also diverge from their own selves from one moment to the next, as God's presence advances with showers of changing moods and circumstances.

Listening and understanding vary with perspectives, experiences, dispositions, and feelings. Exegeses of the Qur'an and the Hadith, the body of the Prophet's sayings, are serious disciplines that devote painstaking attention to the original circumstances of Revelation and recording, chains of transmissions, and authoritative interpretations. Shaykhs are expected to know the exegeses well, and others are expected to remember the limits to their knowledge. Perspectives are irreducibly varied but not equal: the more the perspective can acknowledge and integrate a multitude of vantage points on reality, the better the insight.[20]

Because the Qur'an is at the heart of storytelling, listening presumes training to better hear its message, in whatever imagery tacitly or explicitly evokes it. The poetic quality of Qur'anic expression makes an impression on the reader, but Qur'anic recitation—with its rules, melodies, breaths, and voices that deliberately make sounds, words, and verses stand out—furthers the emotional and visceral experience of listening. Moreover, accomplished storytellers like Shaykh Ayne and his teachers are compelling attention because they embody the message and bring it to life by living and feeling what they tell. They speak to and from a present heart, which is an organ the Qur'an presumes is involved in the reading and listening of the Revelation. *Truly, remembrance of God brings comfort to the hearts.*[21] The following verse elaborates on the felt quality of reception: *God has sent the best statement, a consistent Book, wherein is reiteration. It makes the skins of those who fear their Lord shiver, then their skins and hearts soften at the remembrance of God. Such is God's guidance.*[22]

When Shaykh Ayne and I read al-Ghazālī on death that day, we did not listen to the prose with the same degree of attentiveness. Nor did we listen with the same organs: I listened with my ears and he, presumably, with his skin and heart. He leaned into the chair, shoulders hunched, his breath depressed. He suddenly looked much older and far more frail, his face cloaked in thoughts I dared not prod. Instead, I scrambled to get him to check his blood pressure. Then I fixed him a tea to lift his spirits—a mix of rose, lemon balm, and valerian—and left him to rest.

Our conversations regularly circled back to death and the world's end, and Shaykh's reactions were often strong. Though the composed elder displayed little of his inner states, I could tell that at times, remembrance would strike deep. This capacity of stories to wound and profoundly shake as well as to fill with hope or overwhelm with joy came from the listener's intense, underlying relationship with the Qur'an.

Within Shaykh's world, listening is an art of a living heart.

### *Remembering*

For some Bosnian Muslims, remembrance of death is a deliberate spiritual exercise for cultivating presence, but for most, it is also a part of the quotidian. Personal practices and popular stories play up and improvise on Qur'anic themes to remember the imminent ends.

Five daily prayers, mandatory for Muslims, revolve around the opening chapter of the Qur'an, which one recites in a standing position, praising *God, Lord of the Worlds, the Compassionate, the Merciful, the Master of the Day of Judgment.* The standing during prayer, known as *qiyām*, is meant to foreshadow and prepare one for standing on the Day of Judgment. The end cycle of prayer finds a Muslim sitting, back straight, knees turned forward, eyes turned inward, asking for the good of this world and the good of the next. With that wish, the head turns to the right, invoking peace to the angels on one's right-hand side; the fellow Muslims who are close by in prayer; and, beyond, to everything and everyone in the direction of the head's pointing. Then the head swings to the left, wishing peace on everything in the other direction.

Daily talk spontaneously drifts toward death, one's own, pending, the news of other people's passing, and memories of the dear ones who are deceased. Death is closer than your shirt's collar is a popular folk saying that dresses wisdom into most quotidian garbs.

The subjects of death and kijamet come up in conversation as invitations to shift perspective, to consider life more urgently in light of its passing before we go back to where we were. These are moments of becoming briefly present.

Beekeepers are no exception.

In the summer of 2017, my sister Azra and I have come to the field apiary by the bank of the Sava River for the length of the black locust and false indigo forage to lend a hand to our friends, itinerant beekeepers Sabaheta and Nedžad, with a honey harvest. Nedžad wheels in full honey frames to the field tent where the three of us work, and Sabaheta closely supervises our amateur efforts at uncapping the combs. The wheelbarrow slips through the canvas door swiftly, as we draw the sides of the tent tightly behind us to keep out the bees that trailed the loot.

Taking a break in the tree shade by the tent, in the afternoon, we learn that the couple is in mourning: Sabaheta's first cousin's son has passed away. He was allergic to bee stings, we are told. The last time a bee stung him, he barely survived, so he kept his distance from apiaries ever since and watched his step while outdoors. Sabaheta's cousin worked on the main street at the town butcher. The windows and doors of the shop were kept closed. The air-conditioning was set at temperatures uninvitingly low. Still, a bee wandered in, with no business there—bees are not usually attracted to meat—other than to find him, as our beekeepers put it. It stung him before he noticed that it had landed at the base of his throat, where the soft collar of his shirt parted. The well-built young man, just married, collapsed within minutes, as anaphylaxis seized his body.

"Your death cannot be avoided," Sabaheta says, looking into the distance, her fingers smoothing the edge of her finjan. "The when and how are out of our hands. You go under, empty handed, arms outstretched," she carries on, thinking out loud, "except for your deeds. Their record goes along with you." With this image, we wish God's mercy, "*Raḥmetullāh!*" on the young man and step back into the tent where, within its soft, embryonic walls, honey in combs awaits delivery.

We first met the couple in 2015, just after they had harvested meadow honey in the Kupres highlands in southwest Bosnia. At that time, we learned of their tragic loss during the war. The beekeepers' friends, seated with us at the picnic table, told us what they knew: Sabaheta had been pregnant when their hometown of Žepče came under siege in the 1990s. In a makeshift war hospital, while Nedžad was in the army trenches, Sabaheta gave birth to twins, a boy and a girl. The shelling was constant; the electricity and water supply at the hospital were not. Either at the overwhelmed facility or on the way back from it, something happened, and the newborns' lives were cut short.

The 1990s war provided plenty of opportunities to make finitude an immediately relevant theme, brought home by mass casualties and familial tragedies. War stories, unsurprisingly, circle around death and mortal peril, and the troubled national history since the peace of 1995 often prompts people to describe life as "mere surviving." Remembrance of death is intensified by the experience of precarity, but the everyday rites of remembering have a deeper history and carry Islamic connotations, though to different degrees.

In Western Bosnia, sometime in July 2017, we meet a beekeeper named Hajrudin. Employed as a policeman, Hajrudin spends his free time running a school for beekeeping and the study of the Qur'an with his Muslim friends. Classes for both are offered to students of all ages free of charge and are housed in a modest building by an apiary built on an Islamic land endowment with gifted funds. Honey sales and ongoing gifts sustain the school and pay the teachers' salaries.

Hajrudin is easygoing, his wife and daughters lovely—the younger studies Turkish, the older pharmacy—so we are having a good time together, talking bees and the girls' study plans when our host mentions his dream from the previous night.

> I had a dream last night that I'm preparing my own burial. I'm digging the grave. My brother Muslims come along to give me a hand, and together we dig. It's a fine spot in the cemetery, sort of central. Next, we prepare the boards for my corpse. My deceased father is there, looking on. In the dream, we all know that I will die by the nightfall; minutes go by, I'm parting with the world. And before I die, I wake up thinking the dream is so

true. Any waking day you may go. And that's how a human should live right now. Try and do some *hajr* [good, from the Arabic *khayr*], in whichever form. Because all the rest will come to pass.

## *Returning*

"In the end, everyone returns to God. You can't avoid it," beekeeper Sulejman tells my sister and me a few days later in the same region of Western Bosnia. It was the first time we meet, and my questions during this introductory conversation were largely focused on the details of Sulejman's apicultural practice, and yet the beekeeper's reflections had a wider range. The apiary itself inspired his meditations on finitude.

"I'm not a shaykh, and so I don't bear the grave obligation [to remain mindful at all times], but still, I cannot forget death," Sulejman says. I'm working here, in this near world, but it's passing, and I'm on my way there, to the next, *ākhira*, which is lasting. And the honeybees are here. What I do know, is that bees' song, their collective humming is the sound closest to human *zikr* [invocation]. Truly. Especially in the spring, sometime in the second half of April, when fruit trees bloom, every single hive is sounding the invocations: *Hū-Hū*. The sound is beyond words, the feeling it inspires. Those of us who work around the hives at the time, work in silence. Perhaps it's only in Sufi gatherings that you can experience something similar. A beekeeper becomes hooked on that, rather than on honey, you don't want to lose that, the chance to work with bees.

Sulejman contemplates death in terms of a return to God, which is an ongoing Qur'anic trope. *As He brought you into being, so you shall return*, one verse reads.[23] Or, in the following lines, which Muslims recite for comfort: *But give good tidings to the patient, who say, when a misfortune strikes them, "Indeed, we belong to God, and indeed to Him we shall return."*[24]

The return makes sense of life as a journey away from and back to God, while the point of remembrance is to cultivate the sense that, all along, humans travel in the presence and company of none other than God. That realization ideally ensures that one wakes up and lives more fully before death comes in with the revelation.

The sense of the journey and divine company gives deeper meaning to daily activities—"the honeybees are here." Sulejman hears the bees doing zikr, a loan word from Arabic (*dhikr*) that means "remembrance of God" and commonly refers to individual or collective melodious, immersive, ongoing invocation of God. He hears bees chanting the root of all divine names: *Hū*. Bosnian Muslims who frequent Sufi gatherings commonly observe that honeybees' humming resembles dervishes' voices and breaths joined in devout chanting. A local Sufi lodge plays up these similarities between insects' and humans' devout humming in a yearly performance of what is known as the Bee *dhikr*. Seated on the floor in a ring around Shaykh, dervishes invoke divine names, their voices first slow and soft, then growing stronger and faster, giving the sensation of an airborne bee swarm before coming to settle down again, rippling out in whispers of God's Beautiful Names.

Sulejman's meditations on bees' humming in springtime also suggest that bees help summon the beekeeper into the present and that beekeeping is about keeping up the possibility of a devout, multispecies remembrance. Keeping bees is a way of keeping company with God for whom and with whom the bees buzz adoringly.

Sulejman's life history promoted mindfulness of death and return. In the 1990s, teenage Sulejman, his family, and residents of a small town in central Western Bosnia fled the armed advances of Serb nationalists. Many of his fellow Bosnian Muslims were captured at the time and imprisoned or executed. He returned home five years later when the war ended, carrying hives of honeybees. With that small initial apiary, he rebuilt the devastated homestead and started life anew.

His contemplations about the lifelong process of returning to God keep him from settling into routines too comfortably, though. Thinking eschatologically does not depress one's mood; it is a way of thinking on one's feet, looking forward while pursuing, in the moment, the kinds of values that are everlasting.

### *Rising and Standing*

Bosnian Muslims speak of dying as *preseljenje*, meaning "a relocation," and the vernacular at once tones down the significance of death and imbues the context of dwelling and moving with double significance. The Arabic loan word for this world, *dunja*, "the world close at hand," always resonates with its implicit counterpart, which is "the world after," *ahiret*. For those who were taught to listen to and read the Qur'an, and those who grew up listening to the stories that are the living, growing, shifting thesaurus of the Islamic universe, everyday language suffuses reality with reminders of God. Remembrance is meant to develop one's character, improve one's conduct, promote doing good, and cultivate the sort of knowledge that prepares one for the encounter with God. Sufis presume much the same while emphasizing the inner realm of strivings that aim to enliven one's heart.

The heart's presence with God is a spiritual form of relationship that begins the eschatological process in the here and now. As Shaykh Ayne frequently said: the standing before God takes place now, one's deeds, good and bad, are already swinging the scales, one's death is every moment when the soul is wakeful to the afterlife, the Fire and the Garden open up and close with heartbeats in one's breasts.

The eschatological events after the earth's undoing—a whole series of trials that Islamic textual tradition has elaborated on, building on the hadith and the Qur'an with the aid of contemplation and inspiration—proceed both within the subjective heart reality and outward in resurrected bodies and on the surface of a remade Earth. According to al-Ghazālī, the story goes like this:

God raises to life the angel Isrāfīl, who blows the horn for the second time on the rock of Bayt al-Maqdis in Jerusalem. The tremendous instrument plays up its forty circles of light, the circumference of each like the earth and sky combined, and the radiant holes open up with divine treasure houses until the saved spirits of things "go out with a drone like the droning of bees and fill the entire space from the East and the West. Then, with God's guiding inspiration every soul goes to its body, the beast and the bird and all those creatures having a spirit [*rūh*]."[25]

The spirits are wedded with the bodies, which have grown from some trace—whether forensic or subtle, commentaries diverge—from the earth that was itself brought back to life.[26] Having burned for a thousand years, the charcoal-like planet is rained on, the showers descending from the "sea of life" within divine treasuries until the earth is covered by an ocean and begins to quiver with life.

Al-Ghazālī's account brings us to the point where all beings, "the beast and the bird," not just humans, are raised as witnesses. The proposition can be traced to the following Qur'anic verse. *There is no creature that crawls on the earth, nor bird that flies on its wings, but that they are communities like yourselves—We have neglected nothing in the Book—and they shall be gathered unto their Lord in the end.*[27] Commentators disagree on the question of when the animals are gathered and what their share is in the justice dealt, but the most expansive interpretations on the subject, in keeping with a body of hadith, suggest that animals, too, will be recompensed for their injuries, and their grievances, both against each other and against the humans, will be heard.[28]

Animal witnesses in the eschatological proceedings reiterate the ongoing Qur'anic theme of utter human responsibility for all deeds, as per the following verse: *So whoever does an atom's weight of good will see it, and whoever does an atom's weight of evil will see it.*[29] The verse is emphatic: the animals will be gathered, for nothing is left out of divine concern or divine record. The Book, commentators suggest, may refer to the Qur'an or to what is known as the Preserved Tablet, which keeps a record of all existents, or to the Mother of Books, the source of all revealed scriptures.

The focus of eschatological stories in most accounts, however, remains with the trials of humans for whom the stakes are highest, and who are said to be stunned and left waiting on the day of standing for fifty thousand years, sweating and profoundly uncertain about what will become of them. The length of the wait suggests the tremendousness of the ordeal for the species that the Qur'an describes as hasty and impatient by nature.

Bosnian Muslim references to the End, from patient meditations to passing daily reflections on death, evoke the eschatological proceedings that are waiting on the horizon. Death is the first step toward eternity, and the present is the place of great consequences. It is also the time to live by Revelation, to be enlivened by its reminders of the One, the Living (*al-ḥayy*), who gives life and death and whose name is the Immutable Sustainer (*al-qayyūm*). The Qur'an refers to the Revelation as "Remembrance" (*al-dhikr*) and quotes God as promising: *So remember Me, I will remember you.*[30] To be remembered by God is to be remembered eternally.

Conversely, the Qur'an forewarns, living in forgetfulness earns you divine neglect on Judgment Day: *[God] will say: "So be it. Our revelations came to you and you forgot them, so today you are forgotten."*[31]

Another verse reads: *And be not like those who forgot God and so He made them forget their souls.*[32] The loss, the verses intimate, is utter: one cannot find God the Forgiving if one has forsaken one's soul. There is no self to redeem in the afterlife if the soul, while living, has not found itself through remembering God.

## *Tales of Discomfort*

Islamic eschatological myths, the lore new and old, meditate on and elaborate on the divine promise that *all things perish except His face.*[33] The steadfastness of this promise leaves no room to doubt the coming end, the small and the great, while divine insistence that none other than God knows the timing recommends being on the alert as well as being fully committed to the present, which decides the eternal outcomes. The present in which, the Sufis say, it is possible to die before death so that the heart's eternal life begins without delay.

Eschatological stories ultimately make their listeners uncomfortable. That is their point. For someone like Shaykh Ayne, such discomfort is a sign of accomplished listening skill, while for many Bosnian Muslims, the thought of death or kijamet occasions a sort of brief discomfiting that is manageable and at home in everyday conversations.

The capacity for discomfort is what recommends the image of apocalypse to contemporary thinkers of the global ecological and climate catastrophe. An anthropologist and philosopher renowned in Euro-American academic circles, Bruno Latour recommends the apocalypse as "a prophylactic," a way to better grasp the gravity of climate change projections, given the current trends of anthropogenic warming, in order to organize for countering the catastrophe. In Latour's words: "Fusion of eschatology and ecology is not a plunge into irrationality, a loss of composure, or some sort of mystical adherence to an outdated religious myth; it is a necessity if we want to face up to the threat."[34]

While the sentiment of the statement is rooted in an age-old environmentalist logic—the sheer awareness of catastrophic potential should mobilize public responses—the hasty dismissal of the "outdated religious myths" makes the call to action particularly insensitive toward the plural nature of religious traditions and to the multitude of meanings that are continuously inspired by the old myths. The dismissal intimates a typically modern impatience with things that are "old," the baggage of inheritance that presumably bogs down the imagination. The relevant stories, the assumption goes, are the ones newly invented. The moderns always have the last word.

Katherine Keller is another thinker who has tried to recover the rallying power of the apocalypse. Keller, however, turns to the *Book of Revelation*, the final text of the Christian Bible, and rereads the ancient imagery for fresh insights into current environmental devastation and to imagine probable climate futures, the bleak as well as the hopeful, that might emerge from the rubble. Keller's powerful rereading wrests the tradition from stuffy assumptions and stiff, literalist interpretations, reclaiming it from an irrelevance that is too often ascribed to the scriptural traditions in the context of climate emergency. At the same time, Keller forwards a minimalist metaphysics and polite theology that dulls the most poignant points of Abrahamic tradition, including the notion of God who knows perfectly well what He is doing. Consider the following quote: "We have dremread the text from any recourse to divine control of earth's destiny. Control is not the issue. The issue is what we do, how we live, together. And somehow, sometimes, the Spirit dwells divinely in our togetherness—making it possible. But not making it happen. That's up to us."[35]

The Muslim sources that inform this book, on the contrary, paint the maximal image of God, the *Doer of whatsoever He will*.[36] The Qur'an speaks of God as being closely involved with the world, down to every insect, as well as being most concerned with the intimate affairs of each and every heart, down to its passing moods: *And know that God intervenes between a human and his heart and that to Him you will be gathered*.[37]

This is not a God who kindly lets us live here forever if we so chose but is, on the contrary, the God who promises that everyone will die, as per His plans. The God who anticipates humans' wreaking havoc on the planet, whose existence and presence underwrite all sorts of animacies and activisms, both ruinous and remedial, and, when the time comes, bring about the world's end. This God greatly inconveniences modern human sensibilities, not least our species' presumed sovereignty that is at the heart of Western modern humanism and the environmentally ruthless development, along with the most sacred creed of the supreme human will.

Revelation in Islamic sources revolves around the divine promises that only make sense in the light of finitude and eternity. Imminent death and the world's end reiterate the ecological salience of the tradition, which incites those who care to listen to remember God and stay mindful of human responsibility. Remembrance, *dhikr*, is a thoroughly antiapocalyptic practice that relates humans to the world intensely, tenderly, and warily as the place of searching for the self and God. *Dhikr* is also an ecological practice, as all species of beings, the Book says, save for humans and jinn, unswervingly invoke God.

## *Bees Die like Humans*

I was often told in the field that "bees die like humans" (*pčele umiru k'o ljudi*). My interlocutors pointed out that the vernacular set apart human and apian deaths (*umiru*) from those of other animals (*uginu*). At the level of language, the verb form in common—umiru—makes honeybees and humans somehow kindred in their deaths. I spoke to people who felt that the habit of speech is meaningful, though no one quite knew definitively what species of commonality it implied.

Some Bosnian Muslims thought the saying emphasized a shared sensitivity. "What is not good for the bees is not good for the people either," beekeeper Sulejman suggested. Others speculated that it may have to do with bees' intelligence, language, and social lives, which are uniquely sophisticated among animals. Everyone presumed that the saying referred to the divine revelation, which the Qur'an described as bestowed on both humans and honeybees. The bestowal of Revelation, somehow, makes the two species' lives and deaths inextricably connected and imbues bees with special eschatological significance.

If the honeybees die out, the world will end, beekeepers and bee lovers tell me. The statement is made as a self-evident truth, sometimes presumed to be grounded firmly in Islamic textual tradition. As, for instance, when high-profile businesswoman and hobbyist beekeeper Adisa told me she heard a wise old woman in her neighborhood say: "Forty years after the bees vanish, kijamet will set in." When I press for further details, Adisa shrugs her shoulders, guessing: "She must have read about it in the old books." The expression *old books* (*kitabi*) usually refers to inherited manuscripts in Turkish, Arabic, Arebica (a Bosniak-Persian-Arabic script used during the centuries-long Ottoman period), and Persian, found on the shelves of some Bosnian Muslim family libraries. The reference also presumes that readers are not just well read in multiple languages and cultures of Islam but are also endowed with insights that come from lifelong devotion and contemplation.

Beekeepers, for the most part, blended their apicultural insights and concerns over the bees' growing endangerment along with contemplations of the Qur'an to reason that honeybees' loss portends the end of the world. As Sulejman puts it:

> It is now a challenge to hold onto the bees. We can learn much from them, which is why I often say that we have to serve them well, rather than be their masters. They already have their One Lord. But the human now disturbs them greatly. She [personal pronoun for a bee is she, in BSC] does what she does by means of divine inspiration and will keep doing it until . . . it will be the Day of *Qiyāmah*. We have made such a mess, since we always strive to adapt everything to our own needs. We do it with other beings too. It's just that the honeybees are very sensitive.

Another beekeeper, Imam Hasim, similarly contemplates the human-apian bond. "Bees are a gift from God," Hasim says. "They are inspired by God, and their inspiration should teach us lessons, but, sadly, what do we learn? We are devastating the bees without realizing that by undermining them we are undermining ourselves. Bees are perishing everywhere. The human, *insān*, has now brought them to the edge so they can barely survive the winter."

That honeybees are vital pollinators goes without saying for beekeepers like Hasim and Sulejman, but what they emphasize is the importance of bees' inspiration and the revelatory lessons they hold for humans. When their inspired work ceases, Isrāfīl's horn is bound to sound.

### World without Bees

The endangerment of honeybees has prompted eschatological speculations beyond Bosnia. Amid the ongoing species extinction, it was the dying honeybees, the world's most beloved insect, that incited a particular form of "despair," as noted by environmental philosopher Freya

Mathews.[38] The mass decline of managed colonies from 2005 to 2008, due to the complex environmental, epidemiological, and cultural pressures and glossed as colony collapse disorder (CCD) or colony decline disorder (CDD), inspired blunt news headlines such as "Beepocalypse" and "Bee Armageddon."

Sensationalist headlines aside, the crisis has spurred speculative writing under the theme of the "world without bees." A feature on Greenpeace's website in 2017 opened with the question "Can you imagine a world without bees?" The author promptly answered "I can't" and shifted readers' attention from a bleak future to timely environmental policies, such as the ban on bee-harming pesticides that the article endorsed.[39] Major news media outlets ran stories that pondered what would happen if bees went extinct. Experts debated the extent to which the ongoing decline of managed pollinators threatened the multibillion dollar agricultural and modern food industry and whether a "pollination crisis" could stoke worldwide food insecurity.[40] Popular science books and blogs, apiarists and hobbyists, grassroots activists, and bee-friendly laypeople pondered whether humans would survive the bees' die-off. Another Greenpeace campaign presented an image of a honeybee lying on its back, with the simple caption "If they go so do we." A quote attributed to Albert Einstein, prophesizing that the earth would outlast the loss of bees by a mere four years was widely cited, its oracular appeal unaffected by openly stated doubts about its authenticity or the accuracy of its doom timeline.[41]

Biologists and entomologists strove to put things into perspective, suggesting that the decline of honeybees would not mean the end of the world but merely bleak landscapes and bland diets devoid of staples that make meals tasty and nutritiously rich. These sober accounts sometimes frankly admitted high uncertainty regarding all predictions. In the words of North American biologist Laurence Packer: "We rarely understand the extent to which the continued existence of one species is dependent upon the presence of another."[42] Writing for a general audience, Packer carefully reviewed the great extent to which the commodity-dependent urban existence is contingent on pollination and anticipated that the loss of bees would be catastrophic for biodiversity and food security, with the consequences cascading onwards "through the terrestrial ecosystems of the world."[43]

Fig. 5.1 Being inspired

English-language publications that speculate about a world without bees entertain the proposition that human and apian fates are inseparable and, just as often, backtrack from the boldest implication of the claim to rally people for bee-friendly actions, however low-key, or to have the readers imagine conditions under which humans and their planet would carry on with or without the bees.[44] Recurrent apocalyptic insinuations nonetheless remain attractive and effective, not least in their power to solicit impassioned denunciations.[45]

Popular apocalyptic speculations that spiked in the wake of CCD might have disinclined scientists from researching the implications of climate change on bees. This much was suggested by the science director of the eminent International Bee Research Association (IBRA), Norman Carrack in 2018 interview with *ScieTech Europa Quarterly*. Commenting on a puzzling scarcity of publications in IBRA's journal on the subject of bees and climate change, the director hypothesized: "One of the problems here is perhaps that people have made the wrong argument. For instance, social media often sees a supposed quote from Albert Einstein that if bees disappear then people have around four years to live. There is absolutely no record that Einstein ever said that, and what is more, the statement simply isn't true: most of the world's major staple food crops are wind pollinated."⁴⁶ The insects are crucial for the pollination of crops, especially the foods that contribute variety and nutritional value to the human diet, as well as for the conservation of wild ecosystems, Carrack explains, but they are not indispensable. "The world would not starve without insects," Carrack is quoted saying, "but would be a very dull place."⁴⁷

## At the Beginning

In Bosnian Muslim stories, there is no world without honeybees. I want to take this local proposition at face value and think through its unfinished suggestions on the meaning of honeybees' Revelation and the meaning of divine revelation to the world. Following the statements of my interlocutors, with whom I have been thinking and learning to listen, my reflections are spurred and bound by Islamic tradition, beginning with the following verse in "The Bee" chapter of the Qur'an: *And your Lord revealed to the bee, "Take up dwellings among the mountains and the trees and among that which they build. Then eat of every kind of fruit and follow the ways of Your Nurturer made easy." A drink of various hues comes forth from their bellies wherein there is medicine for humankind. Truly in that is a sign for a people who reflect.*⁴⁸

The Qur'an is taken to be God's last word—"My coming and the coming End are this close," the Prophet of Islam reportedly said, pressing together his index and middle fingers—but the end of the scripture did not end divine revelation. For as long as the world is going, God continues to reveal divine attributes and give news of their work at present through the material, sensual, and conceptual forms that make up the cosmos. What, then, is the Revelation to the world?

A saying attributed to the Prophet much contemplated by the Sufis conveys God's words: "I loved to be known so I brought out the world." This hadith describes cosmogenesis, the world's originating moment or, more specifically, the originating impetus at the heart of it all. In the time before time, when God was Alone in relationship with Self, His desire for self-knowledge initiated the manifestation of prolific, ever-renewed entities and possibilities that the inexhaustive divine reality presumed. The Living One, who is unlike any other, could only manifest as the manyness of singularities. The world is implicitly the world of many knowers, each capable of (and, being engendered by desire, desirous of) receiving the news of God. All species of things, in their different ways and to different degrees, know God and simultaneously broadcast back to the divine self the echoes of multispecies knowing, which comes and returns to the One source. There is no shortchanging a single knower, for each is indispensable, being the very singular way in which God's love of self-knowledge comes about and is met.

While the world continuously reveals and veils God, honeybees are the world's remaining prophetic species.

The bestowal of Revelation, a special form of divine self-disclosure, singles out honeybees and humans: humans because of the heart's capacity—a gift—for getting to know and reflect on all the attributes of divine reality. That, at least, is the ideal to aspire to. The Prophet is the model to emulate, as his mission, a hadith says, is to help Muslims "to perfect character (ākhlaq)," while his character was described as the Qur'an itself. The tradition suggests that taking Revelation to heart is a way for humans to assume and disclose divine qualities. Honeybees live, forage, and dwell by Revelation wholeheartedly; honey flows from it. *Truly in that is a sign for people who reflect.* The fruit of pure inspiration, honey is a viscous, fragrant, sweet form of remembrance, *dhikr*, which is another word for divine revelation.

For those who have learned to listen, honeybees are revelatory. Every revelation, however, presumes another veiling—veil, in the Islamic sense, meaning that which is both a limit to access and the point of intimate contact. Tradition has it that there are seventy thousand veils of light and dark between the world and God. It is because of the veils, Ibn al 'Arabī writes, that we look out and say "This is the world!"[49] If the veils were to be lifted, the lights of divine Being would manifest, and this would no longer be named the world but the God. Veils never lift entirely, for every form of divine self-disclosure—any feeling, thing, being, or idea by which we get to know God better—is also another way in which our comprehension limits the divine reality of God's infinite self-knowledge. Desire is essential to knowing; otherwise, veils would be met with indifference or frustration. In the words of Ibn al 'Arabī, the eyes of the world are enamored with God and are God's lovers in the world, whatever their object of love because all entities are media for God's manifestation.[50]

Honeybees are a special veil, says Zejd, a local dervish with whom I love to think. Death and the End, Zejd writes, are the moments when a veil is lifted between one's heart and God. *You were indeed heedless of this. Now we have removed from you your veil, so today your sight is piercing.*[51] Anything in the meantime that stirs the veil carries an apocalyptic charge of bringing one to presence, refreshing the desire to know the self and God through the world of His beings while there is still time. While the bees are still swarming.

NOTES

1. 40:59.
2. 50:20.
3. 3:9.
4. See Bruce Braun and Sarah J. Whatmore, "The Stuff of Politics: An Introduction," in *Political Matter: Technoscience, Democracy, and Public Life*, ed. Bruce Braun and Sarah J. Whatmore (Minneapolis: University of Minnesota Press, 2010), ix. In an introduction to an insightful volume on elements, editors Jeffrey Cohen and Lowell Duckert make an argument for re-enchanting modern understanding of environments and atmosphere with readings from premodern and early-modern Western sources. The historical sources offer ideas of elements that are capacious, and yet, the editors explicitly state, the theological and metaphysical considerations in their sources are not of interest. John Milton is celebrated for his vitalism while his religious eschatology is left aside. Jeffrey Jerome Cohen and Lowell Duckert, "Introduction: Eleven Principles of the Elements," in *Elemental Ecocriticism: Thinking with Earth, Air, Water, and Fire*, ed. Jeffrey Jerome Cohen and Lowell Duckert (Minneapolis: University of Minnesota Press, 2015), 1–26. Isabelle Stengers, a most adventurous thinker, is interested in the experiences of Virgin Mary pilgrims and the power of nonhumans to make us think and feel, but their theological discussions have no consequence for the analysis of politics that matter. Stengers, "Including Nonhumans in Political Theory," 3–34.
5. Abu Hamid Muhammad Al-Ghazālī, *Ihja' Ulūmid-dīn, Preporod Islamskih Nauka, 8*, trans. Salih Čolaković (Sarajevo: Libris, 2010), 351.
6. Al-Ghazali, *Ihja' Ulūmid-dīn, Preporod Islamskih Nauka, 8*, 370.
7. Ibid., 353, 354.
8. Al-Ghazālī, *Ihja' Ulūmid-dīn, Preporod Islamskih Nauka, 8*, 371.
9. Ibid.
10. 3:185.
11. 75:7–12.
12. 75:26–30.
13. 20:104; 55:37.
14. 40:16.
15. 44:38–39.
16. 79:6–8.
17. Shaykh often mixes Arabic with loan words and Arabic transliterations into BSC. Here, *zikr* is a Bosnian Muslim loan word from Persian via Ottoman Turkish, denoting remembrance and a form of invocation (*dhikr*, in Arabic). *Daimen* is a transliteration of the Arabic adjective denoting "lasting."

18. The appointment letter is written in the *Dīwānī* style of Arabic calligraphy, in which the letters are wavelike and penned in precious ink. The letter sparkles discreetly to this day, on the rare occasions when Shaykh unfolds it.
19. 41:53.
20. Čolić, *Et Tarikatul Muhammedijjetul Islamijjetu*, 14–26.
21. 13:28.
22. 39:23.
23. 7:29.
24. 2:155–156.
25. Abu Hamid Muhammad Al-Ghazali, *The Precious Pearl, Kitab al-Durra al-Fakhira*, trans. Jane Idleman Smith (Missoula, MT: Scholars, 1979), 49. A twentieth-century Sufi, Said Nursi, imagines the resurrection as "the springing to life in an instant of the hundred thousand electric lights of a large city on a festival night, switched on from one center." Said Bediuzzaman Nursi, *Rays: Reflections on Islamic Belief, Thought, Worship and Action*, trans. Hüseyn Akarsu (Clifton, NJ: Tughra, 2010), 35.
26. The issue of resurrection is presented in the Qur'an in the form of an argument for a claim, which, the Book anticipates, sounds outrageous. Voicing the incredulous Arab contemporaries of the Prophet, the Qur'an says: *What! when we become bones and fragments, will we truly be resurrected as a new creation?* And retorts to the mocking challenge with *Say, "Be you stones or iron. Or some other created thing your minds presume still more difficult to raise." And they will say, "Who will restore us?" Say, "He who brought you forth the first time." Then they will nod their heads at you and say, "When is that?" Say, "Perhaps it will be soon."* 17:49–51.
27. 6:38; see also 81:5.
28. A hadith I often heard retold across Bosnia speaks of a woman who earned the Fire for having locked up the cat without food and water until it starved. Shaykh Ayne tells a story that reiterates the Prophet's message with an opposing example. A young, virtuous woman who lived in Ottoman times, Shaykh says, used to spend her wealth building mosques. What earned her the eternal Garden, however, was something else entirely. One day, the story goes, while visiting a mosque construction site, her eye was caught by the sight of an ant stuck in wet mortar. She plucked it out with her little finger. Compassion moves at the smallest twitch, the story suggests. God's Mercy moves it and takes note.
29. 99:7–8.
30. 2:152.
31. 20:126.
32. 59:19.
33. 28:88.
34. Bruno Latour, *Facing Gaia: Eight Lectures on the New Climatic Regime*, trans. Catherine Porter (Cambridge: Polity, 2017), 218.
35. Catherine Keller, *Facing Apocalypse: Climate, Democracy, and Other Last Chances* (Maryknoll, NY: Orbis, 2021), 196.
36. 85:16.
37. 8:24.
38. Mathews starts with the powerful hold honeybees have over "us"—her presumed modern readers—to rethink "who is the honeybee, and what have we done to her?" While premodern cultures explored such questions with cosmological stories, the moderns seek answers by means of science, Mathews writes, because science, "for better or worse," is the modern way of making sense of the world. Mathews's ambivalence arises from her recognition that the "scientific approach may be part of what has led to the honeybee's endangerment," presumably through disenchantment but also through apicultural practices that made honeybees a workforce of industrial agriculture, which she resolves with philosophical speculation on hive consciousness and deep resonance between the planet and the beehive, the work of pollination and the metabolic synergies that keep up the biosphere. Mathews deliberately steers inquiry

beyond the instrumentalist arguments about the value of the pollination service and shows that even ethical consideration fails to match the affective surplus that bees' plight generates for a modern audience. Freya Mathews, "Planet Beehive," *Australian Humanities Review* 50 (May 2011), http://australianhumanitiesreview.org/2011/05/01/planet-beehive/.

39. "Politicians Need to Hear Our Buzz and Act," Luis Ferreirim, "Can You Imagine a World without Bees?," Greenpeace, Stories, Nature, April 27, 2017, https://www.greenpeace.org/international/story/7578/can-you-imagine-a-world-without-bees/. See also UNFAO (Food and Agricultural Organization of the United Nations), "Imagine a World without Bees," UNFAO YouTube channel, video, 01:26, posted May 18, 2018, https://www.youtube.com/watch?v=el-Z5tgyQXY.

40. Honeybees' decline has been framed as a pollination crisis, which is essentially a threat to the world's food security. In 2006, J. Ghazoul wrote an admittedly skeptical response to that claim, arguing that insect pollinators are marginal to food production, that the declines were confined to industrially employed bees and bumblebees in North America and Europe and that there is insufficient evidence for concern that anthropogenic factors lead to insect declines. Jaboury Ghazoul, "Buzziness as Usual? Questioning the Global Pollination Crisis," *Trends in Ecology & Evolution* 20, no. 7 (July 2005): 367–373. A comprehensive response was issued in 2010 by Potts et al., which reinstated the importance of pollinators for food security and biodiversity, arguing that 75 percent of all crops used directly for human food worldwide depend on insect pollinators, and 80 percent of wild plants depend on the insects for their production of fruits and flowers. They also presented a wider range of evidence for the loss of diversity and the abundance of wild bees in the United Kingdom and the Netherlands and parallel losses in plant communities dependent on pollinators. The bulk of evidence from metastudies shows a widespread pattern of loss of pollinator diversity and abundance as a result of agricultural intensification and habitat loss, and since most natural landscapes around the world are anthropogenically modified, the authors suggest that the losses are likely to be ongoing around the world. Simon G. Potts et al., "Global Pollinator Declines: Trends, Impacts and Drivers," *Trends in Ecology & Evolution* 25, no. 6 (February 2010): 350, https://doi.org/10.1016/j.tree.2010.01.007.

41. See, for instance, Allison Benjamin and Brian McCallum, *A World without Bees* (New York: Pegasus, 2009), 7.

42. Laurence Packer, *Keeping the Bees: Why All Bees Are at Risk and What We Can Do to Save Them* (Toronto: HarperCollins, 2010), 5.

43. Ibid., 4–5.

44. See Marla Spivak, "What Will Happen If the Bees Disappear?," CNN, March 6, 2015, https://edition.cnn.com/2014/05/17/opinion/spivak-loss-of-bees/index.html.

45. See Christopher Ingraham, "Call Off the Bee-Pocalypse: U.S. Honeybee Colonies Hit a 20-Year High," *Washington Post*, July 23, 2015, https://www.washingtonpost.com/news/wonk/wp/2015/07/23/call-off-the-bee-pocalypse-u-s-honeybee-colonies-hit-a-20-year-high/; Shawn Regan, "What Happened to the 'Bee-pocalypse'? A New Study Explores How Pollination Markets Saved the Bees," PERC, posted July 12, 2019, https://www.perc.org/2019/07/12/what-happened-to-the-bee-pocalypse/.

46. *Scitech Europa Quarterly*, "Bees and the Changing Climate," Environment & Sustainability News, accessed February 16, 2018, https://www.scitecheuropa.eu/bee-populations-changing-climate/84417/.

47. Ibid.

48. 16:68–69.

49. Ibn al-'Arabi, *Mekanska oktrovenja*, 695.

50. Ibid., 475.

51. 50:22.

# BIBLIOGRAPHY

Abadžić, Nijaz. *Tajne pčelinjeg meda*. Sarajevo: NIP "Zadrugar," 1967.
Alaimo, Stacey. *Bodily Natures: Science, Environment, and the Material Stuff*. Bloomington: Indiana University Press, 2010.
———. "Elemental Love in the Anthropocene." In *Elemental Ecocriticism: Thinking with Earth, Air, Water, and Fire*, edited by Jeffrey Jerome Cohen and Lowell Duckert, 298–309. Minneapolis: University of Minnesota Press, 2015.
———. *Exposed: Environmental Politics and Pleasures in Posthuman Times*. Minneapolis: University of Minnesota Press, 2016.
Alaux, Cédric, Cristelle Dantec, Hughes Parrinello, and Yves Le Conte. "Nutrigenomics in Honey Bees: Digital Gene Expression Analysis of Pollen's Nutritive Effects on Healthy and Varroa-Parasitizes Bees." *BMC Genomics* 12 (October 2011): 496. https://doi.org/10.1186/1471-2164-12-496.
Al-Ghazali, Abu Hamid Muhammad. *Ihja' Ulūmid-dīn, Preporod Islamskih Nauka, 2*. Translated by Salih Čolaković. Sarajevo: Libris, 2010.
———. *Ihja' Ulūmid-dīn, Preporod Islamskih Nauka, 7*. Translated by Salih Čolaković. Sarajevo: Libris, 2010.
———. *Ihja' Ulūmid-dīn, Preporod Islamskih Nauka, 8*. Translated by Salih Čolaković. Sarajevo: Libris, 2010.
———. *The Precious Pearl Al-Durra Al-Fakhira*. Translated by Jane Idleman Smith. Missoula: Scholars, 1979.
———. *The Remembrance of Death and the Afterlife*. Translated by T. J. Winter. Cambridge: Islamic Texts Society, 1989.

Al Jazeera Balkans. "Teška godina za pčelare u regionu." *Al Jazeera Balkans*, July 30, 2016. http://balkans.aljazeera.net/vijesti/teska-godina-za-pcelare-u-regionu.

Al-Samie, Abdul Rahim, and Mohamed Ali. "Studies on Bee Venom and Its Medical Uses." *International Journal of Advancements in Research & Technologie* 1, no. 2 (July 2012). https://pdfs.semanticscholar.org/a0b1/f2ed65fc1b999b657045135d563 e7bfd298b.pdf?_ga=2.55221038.485531400.1581172971-2113641594.1577775850.

Audit Office for the Institutions of the Federation BiH. "Izvještaj revizije učinka prevencija poplava u Federaciji BiH, 2013." Performance Audit. Posted January 21, 2013. https://www.vrifbih.ba/?s=Izvještaj+revizije+učinka+prevencija+poplava+u+Federaciji+BiH,+2013&post_type=.

Benjamin, Allison, and Brian McCallum. *A World without Bees*. New York: Pegasus, 2009.

Bennett, Jane. *Thoreau's Nature: Ethics, Politics, and the Wild*. Lanham, MD: Rowman & Littlefield, 2002.

BH Pčelar. "Pogledi u nebo i molba Svevisnjem, Sulejman Alijagic." *BH Pčelar* 36, June 15–August 15, 2014.

———. "Priroda da se okrenula protiv nas ili mi protiv nje? Nista ne bi od 'zlatne godine,' Rajko Radivojac." *BH Pčelar* 36, June 15–August 15, 2014.

Blok, Anders. "War of the Whales: Post-Sovereign Science and Agnostic Cosmopolitics in Japanese-Global Whaling Assemblages." *Science, Technology & Human Values* 36, no. 1 (November 2010): 55–81. https://doi.org/10.1177/0162243910366133.

BN televizija. "Nema elementarne nepogode?!" *BN televizija*, August 11, 2017. https://www.rtvbn.com/3875127/nema-elementarne-nepogode.

Bogdanov, Stefan, Tomislav Jurendic, Robert Sieber, and Peter Gallman. "Honey for Nutrition and Health: A Review." *Journal of the American College of Nutrition* 27, no. 6 (December 2008): 677–689. https://doi.org/10.1080/07315724.2008.10719745.

Braun, Bruce, and Sarah J. Whatmore. "The Stuff of Politics: An Introduction." In *Political Matter: Technoscience, Democracy, and Public Life*, edited by Bruce Braun and Sarah J. Whatmore, ix. Minneapolis: University of Minnesota Press, 2010.

Čabro, Rahela. "Teško ljeto za poljoprivrednike u BiH, suša uništila veliki broj usjeva." TNT portal, August 11, 2017. https://tntportal.ba/vijesti/tesko-ljeto-za-poljoprivrednike-u-bih-susa-unistila-veliki-broj-usjeva/.

Caldwell, Melissa. *Dacha Idylls: Living Organically in Russia's Countryside*. Berkeley: University of California Press, 2010.

Center for Climate and Energy Solutions. "Heat Waves and Climate Change." Climate Basics. Extreme Weather. Accessed October 17, 2019. https://www.c2es.org/content/heat-waves-and-climate-change/.

Center for Ecology and Energy Tuzla. "Deterdženti bez fosfata—napredak za okoliš." Projects. Posted January 2012. https://ekologija.ba/2017/05/22/deterdzenti-bez-fosfata/.

Chen, Jun, Su-Min Guan, Wei Sun, and Han Fu. "Melittin, the Major Pain-Producing Substance of Bee Venom." *Neuroscience Bulletin* 32 (March 2016): 265–272. https://doi.org/10.1007/s12264-016-0024-y.

Chen, Mel. *Animacies: Biopolitics, Racial Mattering, and Queer Affect*. Durham, NC: Duke University Press, 2012.

Chittick, William. *Divine Love: Islamic Literature and the Path to God*. New Haven, CT: Yale University Press, 2013.

———. "Eschatology." In *Islamic Spirituality: Foundations*, edited by Seyyed Hossein Nasr, 378–409. New York: Crossroad, 1997.

———. "Muslim Eschatology." In *The Oxford Handbook of Eschatology*, edited by Jerry L. Walls, 132–150. Oxford: Oxford University Press, 2008.

———. *The Sufi Path of Knowledge: Ibn al Arabi's Metaphysics of Imagination*. Albany: State University of New York Press, 1989.

Coates, Peter. *Ibn 'Arabi and Modern Thought. The History of Taking Metaphysics Seriously*. Oxford: Anqa Publishing, 2011.

Cohen, Jeffrey Jerome, and Lowell Duckert. "Introduction: Eleven Principles of the Elements." In *Elemental Ecocriticism: Thinking with Earth, Air, Water, and Fire*, edited by Jeffrey Jerome Cohen and Lowell Duckert, 1–26. Minneapolis: University of Minnesota Press, 2015.

Čolić, Mustafa. *Et Tarikatul Muhammedijjetul Islamijjetu: Evidencije i definicije islamskih šerijatskih učenja i vjerovanja*. Visoko: Kaligraf, 1998.

———. *Et-Tarikatul-Muhammedijjetul-Islamijjetu, učenje i moral Allahovog Poslanika Muhammeda a.s.: Srčano zdravlje i bolesti metafizičkog insana*. Visoko: Tekija Šejh Husejn-baba Zukić-Zivčići, 2016.

———. *Et Tarikatul Muhammedijjetul Islamijjetu: Učenje i moral Allahovog Poslanika Muhammeda a.s.: Zdravlje i bolesti jezika i ostalih organa metafizičkog insana*. Visoko: Tekija Šejh Husejn-baba Zukić-Živčići, 2020.

———. *Kelamske i Tekvinske Božanske Obznambene Objave i Pojave i Njihovi Kira'eti (Čitanje i Učenje)*. Visoko: Tekija Šejh Husejn-baba Zukić-Živčići, 2003.

———. *Zagonetnosti i Nepoznanice Metafizičkog Srca (Metafizičkog Čovjeka)*. Visoko: Tekija Šejh Husejn-baba Zukić-Zivčići, 2000.

———. *Zbirni Ilmihal islamizacionih stanja i ptanja za odrasle i dorasle*. Visoko: Tekija Šejh Husejn-baba Zukić-Živčići, 2000.

Cook, David. *Contemporary Muslim Apocalyptic Literature*. Syracuse: Syracuse University Press, 2008.

Crane, Eva. *A Book of Honey*. Oxford: Oxford University Press, 1980.

da Silva, Elizamar Ciríaco, Manoel Albuquerque, André Dias, André Azevedo Neto, and Carlos Dias da Silva Junior. "Drought and Its Consequences to Plants—From Individual to Ecosystem." In *Responses of Organisms to Water Stress*, edited by Sener Akinci, 17–47. London: IntechOpen, 2013.

Deane-Drummond, Celia, and David Clough, eds. *Creaturely Theology: On God, Humans and Other Animals*. London: SCM, 2009.

de la Cadena, Marisol. "Indigenous Cosmopolitics in the Andes: Conceptual Reflections beyond 'Politics.'" *Cultural Anthropology* 25, no. 2 (April 2010): 334–370. https://doi.org/10.1111/j.1548-1360.2010.01061.x.

Demos, T. J. *Decolonizing Nature: Contemporary Art and the Politics of Ecology*. Berlin: Sternberg, 2016.

Dibley, Ben. "Anthropocene: The Enigma of 'The Geomorphic Fold.'" In *Animals in the Anthropocene: Critical Perspectives on Non-Human Futures*, edited by Human

Animal Research Network Editorial Collective, 19–32. Sydney: Sydney University Press, 2015.

Dien, Mawil Izzi. *The Environmental Dimensions of Islam*. Cambridge: Lutterworth, 2000.

Easton-Calabria, August, Kristian C. Demary, and Nola J. Oner. "Beyond Pollination: Honey Bees (*Apis mellifera*) as Zootherapy Keystone Species." *Frontiers in Ecology and Evolution* 6 (February 2019): 161. https://doi.org/10.3389/fevo.2018.00161.

Eteraf-Oskouei, Tahereh, and Moslem Najafi. "Traditional and Modern Uses of Honey in Human Diseases: A Review." *Iranian Journal of Basic Medical Science* 16, no. 6 (June 2013): 731–742. https://doi.org/10.22038/IJBMS.2013.988.

Eze, Obioma, Okwesili Nwodo, and Victor Ogugua. "Therapeutic Effect of Honey Bee Venom." *Journal of Pharmaceutical, Chemical, and Biological Sciences* 4, no. 1 (March–May 2016): 48–53. https://www.semanticscholar.org/paper/Therapeutic-Effect-of-Honey-Bee-Venom-Eze-Nwodo/4be85dc94738742f9102b2694273835d57a8ceb2.

Fakhar-i-Abbas. *Animals Rights in Islam: Islam and Animal's Rights*. Riga: VDM Verlag, 2009.

FENA. "SUŠA DRASTIČNO SMANJILA PROIZVODNJU MEDA: Tržište će da preplavi VEŠTAČKI." *Blic*, August 16, 2017. https://www.blic.rs/vesti/republika-srpska/susa-drasticno-smanjila-proizvodnju-meda-trziste-ce-da-preplavi-vestacki/wjqmf6k.

Ferreirim, Luis. "Can You Imagine a World without Bees?" Greenpeace. Stories. Nature. Posted April 27, 2017. https://www.greenpeace.org/international/story/7578/can-you-imagine-a-world-without-bees/.

Filiu, Jean-Pierre. *Apocalypse in Islam*. Translated by M. B. DeBevoise. Oakland: University of California Press, 2011.

Foltz, C. Richard, Frederick M. Denny, and Azizan Baharuddin, eds. *Islam and Ecology: A Bestowed Trust*. Cambridge, MA: Center for the Study of World Religions, Harvard Divinity School, 2003.

Foltz, Richard. *Animals in Islamic Tradition and Muslim Cultures*. London: Oneworld, 2005.

———. "'This She-Camel of God Is a Sign to You': Dimensions of Animals in Islamic Tradition and Muslim Culture." In *A Communion of Subjects: Animals in Religion, Science, and Ethics*, edited by Paul Waldau and Kimberly Patton, 149–159. New York: Columbia University Press, 2009.

Foster, John. *After Sustainability: Denial, Hope, Retrieval*. New York: Routledge, 2015.

Foucault, Michel. *Politics, Philosophy, Culture: Interviews and Other Writings 1977–1984*. Translated by Alan Sheridan. New York: Routledge, Chapman & Hall, 1988.

Fowler, Hayden. "Epilogue New World Order—Nature in the Anthropocene." In *Animals in the Anthropocene: Critical Perspectives on Non-Human Futures*, edited by Human Animal Research Network Editorial Collective, 243–254. Sydney: Sydney University Press, 2015.

Gauldie, Jack, Jennifer M. Hanson, Franklin D. Rumjanek, Rudolf A. Shipolini, and Charles A. Vernon. "The Peptide Components of Bee Venom." *European Journal of Biochemistry* 61, no. 2 (January 1976): 369–376. https://doi.org/10.1111/j.1432-1033.1976.tb10030.x.

Ghazoul, Jaboury. "Buzziness as Usual? Questioning the Global Pollination Crisis." *Trends in Ecology & Evolution* 20, no. 7 (July 2005): 367–373. https://doi.org/10.1016/j.tree.2005.04.026.

Guarasci, Bridget, and Eleana J. Kim. "Ecologies of War." Theorizing Contemporary. Fieldsights. Posted January 25, 2022. https://culanth.org/fieldsights/series/ecologies-of-war.

Günther, Sebastian, and Todd Lawson, eds. *Roads to Paradise: Eschatology and Concepts of the Hereafter in Islam*. Leiden: Brill, 2017.

Gupta, Rakesh Kumar, and Stefan Stangaciu. "Apitherapy." In *Beekeeping for Poverty Alleviation and Livelihood Security*, edited by Rakesh Gupta, Wim Reybroeck, Johan W. van Veen, and Anuradha Gupta, 413–446. Dordrecht: Springer, 2014.

H. "Lucifer poharao Europu: najmanje dvoje ljudi umrlo od vrućine, najteže pogođeni Balkan i Italija, na Jadranu i danas paklenih 42 stupnja." *Slobodna Dalmacija*, August 4, 2017. https://slobodnadalmacija.hr/vijesti/svijet/lucifer-poharao-europu-najmanje-dvoje-ljudi-umrlo-od-vrucine-najteze-pogodeni-balkan-i-italija-na-jadranu-i-danas-paklenih-42-stupnja-500315.

Hamilton, Clive. *Requiem for a Species: Why We Resist the Truth about Climate Change*. New York: Earthscan, 2015.

Haraway, Donna. "Otherworldly Conversations, Terran Topics, Local Terms." In *Material Feminisms*, edited by Stacey Alaimo and Susan Hekman, 157–187. Bloomington: Indiana University Press, 2008.

———. *When Species Meet*. Minneapolis: University of Minnesota Press, 2008.

Hegland, Joar Stein, Anders Nielsen, Amparo Lazaro, Ann-Line Bjerkens, and Orjan Totland. "How Does Climate Warming Affect Plant-Pollinator Interactions?" *Ecology Letters* 12, no. 2 (February 2009): 184–195. https://doi.org/10.1111/j.1461-0248.2008.01269.x.

Henig, David. *Remaking Bosnian Muslim Lives: Everyday Islam in Postwar Bosnia and Herzegovina*. Champaign: University of Illinois Press, 2020.

Human Animal Research Network Editorial Collective. *Animals in the Anthropocene: Critical Perspectives on Non-Human Futures*. Sydney: Sydney University Press, 2015.

Ibn al-'Arabi, Muhyiddin. *Mekanska otkrovenja*. Translated by Salih Ibrišević and Ismail Ahmetagić. Sarajevo: Ibn Arebi, 2007.

Ibn Kathīr, Ismail. *The Signs before the Day of Judgment*. London: Dar Al Taqwa, 1997.

ILO (International Labor Organization). "Bosnia and Herzegovina Floods 2014: Recovery Needs Assessment." Report. Posted June 30, 2014. https://www.ilo.org/global/topics/employment-promotion/recovery-and-reconstruction/WCMS_397687/lang--en/index.htm.

Ingraham, Christopher. "Call Off the Bee-Pocalypse: U.S. Honeybee Colonies Hit a 20-Year High." *Washington Post*, July 23, 2015. https://www.washingtonpost.com/news/wonk/wp/2015/07/23/call-off-the-bee-pocalypse-u-s-honeybee-colonies-hit-a-20-year-high/.

Jaworski, Coline, Benoît Geslin, and Catherine Fernandez. "Climate Change: Bees Are Disoriented by Flowers' Changing Scents." *The Conversation*, June 25, 2019.

http://theconversation.com/climate-change-bees-are-disorientated-by-flowers-changing-scents-119256.

Jojriš, Naum Petrović. *Pčele i medicina*. Translated by Aleksandar Đerdanović. Banja Luka: Romanov, 1974.

Jones, Julia C., Paul Helliwell, Madeleine Beekman, Ryszard Maleszka, and Benjamin P. Oldroyd. "The Effects of Rearing Temperature on Developmental Stability and Learning and Memory in the Honey Bee, *Apis mellifera*." *Journal of Comparative Physiology A* 191 (December 2005): 1121–1129. https://doi.org/10.1007/s00359-005-0035-z.

Keller, Catherine. *Facing Apocalypse: Climate, Democracy, and Other Last Chances*. Maryknoll, NY: Orbis, 2021.

Kendić, Armin. "Kruks, živa i hlor u industrijskoj zoni: Opasan otpad prijeti zdravlju gradjana Tuzle." *Klix*, March 5, 2016. https://www.klix.ba/vijesti/bih/kruks-ziva-i-hlor-u-industrijskoj-zoni-opasan-otpad-prijeti-zdravlju-gradjana-tuzle/160304151.

Khazaei, Mozafar, Atefe Ansarian, and Elham Ghanbari. "New Findings on Biological Actions and Clinical Applications of Royal Jelly." *Journal of Dietary Supplements* 15, no. 3 (October 2017): 1–19. https://doi.org/10.1080/19390211.2017.1363843.

Kirksey, Eben. *Emergent Ecologies*. Durham, NC: Duke University Press, 2015.

Korkut, Besim. *Kur'an s prijevodom*. Medina: Hadimu-l-Haramejni-š-Šerifejni-l-Melik Fahd, 1992.

Kuropatnicki, Andrzej, Ewelina Szliszka, and Wojciech Kroll. "Historical Aspects of Propolis Research in Modern Times." *Evidence-Based Complementary and Alternative Medicine* 2013 (2013): 964149. http://dx.doi.org/10.1155/2013/964149.

Langwick, Stacey Ann. *Bodies, Politics, and African Healing: The Matter of Maladies in Tanzania*. Bloomington: Indiana University Press, 2011.

Latour, Bruno. *Facing Gaia: Eight Lectures on the New Climatic Regime*. Translated by Catherine Porter. Cambridge: Polity, 2017.

———. *An Inquiry into Modes of Existence: An Anthropology of the Moderns*. Translated by Catherine Porter. Harvard, MA: Harvard University Press, 2013.

———. *Politics of Nature: How to Bring the Sciences into Democracy*. Translated by Catherine Porter. Harvard, MA: Harvard University Press, 2004.

Le Conte, Yves, and Maria Navajas. "Climate Change: Impact on Honey Bee Populations and Diseases." *Revue scinetifique et technique* 27, no. 7 (August 2008): 499–510. http://dx.doi.org/10.20506/rst.27.2.1819.

Lemery, Jay, and Paul Auerbach. *The Enviromedics: The Impact of Climate Change on Human Health*. Lanham, MD: Rowman & Littlefield, 2017.

Lorimer, Jamie. *Wildlife in the Anthropocene: Conservation after Nature*. Minneapolis: University of Minnesota Press, 2015.

Marquet, Pablo, Janeth Lessmann, and M. Rebecca Shaw. "Protected-Area Management and Climate Change." In *Biodiversity and Climate Change: Transforming the Biosphere*, edited by Thomas E. Lovejoy, Lee Hannah, and Edward O. Wilson, 283–293. New Haven, CT: Yale University Press, 2019.

Masri, Al-Hafiz Basheer Ahmad. *Animal Welfare in Islam*. Markfield: Islamic Foundation, 2007.

Mathews, Freya. "Planet Beehive." *Australian Humanities Review* 50 (May 2011). http://australianhumanitiesreview.org/2011/05/01/planet-beehive/.

McKibben, Bill. *The End of Nature*. New York: Random House Trade Paperbacks, 2006.

Miller, Adam S. *Speculative Grace: Bruno Latour and Object-Oriented Theology*. New York: Fordham University Press, 2013.

Ministry of Security of Bosnia and Herzegovina. "Procjena ugroženosti Bosne i Hercegovine od prirodnih ili drugih nesreća, 2011." Documents. Other Documents. Posted March 31, 2014. http://www.msb.gov.ba/Zakoni/dokumenti/default.aspx?id=10773&langTag=bs-BA.

Mirzoeva, Olga, Ruslan Grishanin, and Philip Calder. "Antimicrobial Action of Propolis and Some of Its Components: The Effects on Growth, Membrane Potential and Motility of Bacteria." *Microbiological Research* 152, no. 3 (September 1997): 239–246. https://doi.org/10.1016/S0944-5013(97)80034-1.

Molan, Peter. "Why Honey Is Effective as a Medicine. 1. Its Use in Modern Medicine." *Bee World* 80, no. 2 (1999): 80–92. https://hdl.handle.net/10289/2059.

Moore, Stephen, ed. *Divinanimality: Animal Theory, Creaturely Theology*. New York: Fordham University Press, 2014.

Morton, Timothy. *Ecology without Nature: Rethinking Environmental Aesthetics*. Cambridge, MA: Harvard University Press, 2007.

Murata, Sachiko. *The Tao of Islam: A Sourcebook on Gender Relationships in Islamic Thought*. New York: State University of New York Press, 1992.

Nancy, Jean-Luc. *After Fukushima: The Equivalence of Catastrophes*. Translated by Charlotte Mandell. New York: Fordham University, 2015.

Nasr, Seyyed Hossein. *Knowledge and the Sacred*. Albany: State University of New York Press, 1989.

———. *Man and Nature: The Spiritual Crisis in Modern Man*. Chicago: Kazi, 2003.

Nasr, Seyyed Hossein, Caner K. Dagli, Maria Massi Dakake, Joseph E. B. Lumbard, and Mohammed Rustom. *The Study Qur'an: A New Translation and Commentary*. New York: HarperCollins, 2017.

Nursi, Said Bediuzzaman. *The Flashes Collection*. Translated by Şükran Vahide. Istanbul: Reyan Ofset, 2002.

———. *Rays: Reflections on Islamic Belief, Thought, Worship and Action*. Translated by Hüseyn Akarsu. Clifton, NJ: Tughra, 2010.

Ornella, Lai Moon Dor Ginnie, and Fawzi Mahomoodally. "Traditional and Modern Uses of Honey: An Updated Review." In *Honey: Geographical Origins, Bioactive Properties and Health Benefits*, edited by Ruben Ramirez, 81–98. New York: Nova Science, 2016.

Pacini, Ettore, and Massimo Nepi. "Nectar Production and Presentation." In *Nectaries and Nectar*, edited by Susan W. Nicolson, Massimo Nepi, and Ettore Pacini, 167–214. Dordrecht: Springer, 2017.

Packer, Laurence. *Keeping the Bees: Why All Bees Are at Risk and What We Can Do to Save Them*. Toronto: HarperCollins, 2010.

Parmesan, Camille. "Influences of Species, Latitudes and Methodologies on Estimates of Phenological Response to Global Warming." *Global Change Biology* 13 (August 2007): 1860–1872. https://doi.org/10.1111/j.1365-2486.2007.01404.x.

Pasupuleti, Visweswara Rao, Lakhsmi Sammugam, Nagesvari Ramesh, and Siew Hua Gan. "Honey, Propolis, and Royal Jelly: A Comprehensive Review of Their Biological Actions and Health Benefits." *Oxidative Medicine and Cellular Longevity* 2017 (2017): 1259510, https://doi.org/10.1155/2017/1259510.

Petanidou, Theodora, and Erik Smets. "Does Temperature Stress Induce Nectar Secretion in Mediterranean Plants?" *New Phytologist* 133, no. 3 (July 1996): 513–518. https://doi.org/10.1111/j.1469-8137.1996.tb01919.x.

Pick, Anat. *Creaturely Poetics: Animality and Vulnerability in Literature and Film*. New York: Columbia University Press, 2011.

Post, Eric, and Michael Avery. "Phenological Dynamics in Pollinator-Plant Associations Related to Climate Change." In *Biodiversity and Climate Change: Transforming the Biosphere*, edited by Thomas E. Lovejoy and Lee Hannah, 42–54. New Haven, CT: Yale University Press, 2019.

Potts, G. Simon, Jacobus C. Biesmeijer, Claire Kremen, Peter Neumann, Oliver Schweiger, and William E. Kunin. "Global Pollinator Declines: Trends, Impacts and Drivers." *Trends in Ecology & Evolution* 25, no. 6 (February 2010): 345–353. https://doi.org/10.1016/j.tree.2010.01.007.

Prado, Alberto, Fabrice Requier, Didier Crauser, Yves Le Conte, Vincent Bretagnolle, and Cédric Alaux. "Honeybee Lifespan: The Critical Role of Pre-Foraging Stage." *Royal Society Open Science* 7 (October 2020): 200998. https://royalsocietypublishing.org/doi/pdf/10.1098/rsos.200998.

Regan, Shawn. "What Happened to the 'Bee-pocalypse'? A New Study Explores How Pollination Markets Saved the Bees." PERC. Posted July 12, 2019. https://www.perc.org/2019/07/12/what-happened-to-the-bee-pocalypse/.

Root, Amos Ives. *The ABC & XYZ of Bee Culture*. Medina, OH: A. I. Root, 2007.

Rose, Deborah Bird. *Wild Dog Dreaming: Love and Extinction*. Charlottesville: University of Virginia Press, 2011.

Rueppell, Olav, Osman Kaftanouglu, and Robert E. Page Jr. "Honey Bee (Apis mellifera) Workers Live Longer in Small than in Large Colonies." *Experimental Gerontology* 44 (April 2009): 447–452. https://doi.org/10.1016/j.exger.2009.04.003.

Saltykova, Elena, A. A. Karimova, A. R. Gataullin, Louisa Gaifullina, Rustam Matniyazov, A.I. Albulov, and Alexey Nikolenko. "The Effect of High-Molecular Weight Chitosans on the Antioxidant and Immune Systems of the Honeybee." *Applied Biochemistry and Microbiology* 52, no. 5 (September 2016): 553–557. https://doi.org/10.1134/S0003683816050136.

Scaven, Victoria, and Nicole E. Rafferty. "Physiological Effects of Climate Warming on Flowering Plants and Insect Pollinators and Potential Consequences for Their Interactions." *Current Zoology* 59, no. 3 (September 2013): 418–426. https://doi.org/10.1093/czoolo/59.3.418.

Scitech Europa. "Bees and the Changing Climate." Environment & Sustainability News. Accessed February 16, 2018. https://www.scitecheuropa.eu/bee-populations-changing-climate/84417/.

Seeley, Thomas Dyer. *Honeybee Democracy*. Princeton, NJ: Princeton University Press, 2010.

———. *The Lives of Bees: The Untold Story of the Honey Bee in the Wild*. Princeton, NJ: Princeton University Press, 2019.

Skafish, Peter. "Anthropological Metaphysics/Philosophical Resistance." Theorizing Contemporary. Fieldsights. Posted January 13, 2014. https://culanth.org/fieldsights/anthropological-metaphysics-philosophical-resistance.

Smith, Jane Idelman, and Yvonne Yazbeck Haddad. *The Islamic Understanding of Death and Resurrection*. Oxford: Oxford University Press, 2002.

Soper, Kate. *What Is Nature?* Malden, MA: Blackwell, 1995.

Spivak, Marla. "What Will Happen If the Bees Disappear?" CNN, March 6, 2015, https://edition.cnn.com/2014/05/17/opinion/spivak-loss-of-bees/index.html.

Stengers, Isabelle. *Cosmopolitics 1*. Translated by Robert Bononno. Minneapolis: University of Minnesota Press, 2010.

———. "Including Nonhumans in Political Theory: Opening Pandora's Box?" In *Political Matter: Technoscience, Democracy, and Public Life*, edited by Bruce Braun and Sarah J. Whatmore, 3–34. Minneapolis: University of Minnesota Press, 2010.

———. *In the Catastrophic Times: Resisting the Coming Barbarism*. Translated by Andrew Goffey. London: Open Humanities Press with Meson Press, 2015.

———. *The Invention of Modern Science*. Translated by Daniel W. Smith. Minneapolis: University of Minnesota Press, 2000.

———. "Reclaiming Animism." *e-flux journal* 36 (July 2012): 1–12. https://www.e-flux.com/journal/36/61245/reclaiming-animism/.

———. "Wondering about Materialism." In *The Speculative Turn: Continental Materialism and Realism*, edited by Levi Bryant, Nick Srnicek, and Graham Harman, 368–380. Melbourne: re.press, 2011.

Takkis, Krista, Thomas Tscheulin, Panagiotis Tsalkatis, and Theodora Petanidou. "Climate Change Reduces Nectar Secretion in Two Common Mediterranean Plants." *AoB Plants* 7 (September 2015): plv111. https://doi.org/10.1093/aobpla/plv111.

Tan, K., F. Bock, S. Fuchs, S. Streit, A. Brockman, and J. Tautz. "Effects of Brood Temperature on Honey Bee *Apis mellifera* Wing Morphology." *Acta Zoologica Sinica* 51, no. 4 (2005): 768–771.

Tautz, Jürgen. *The Buzz about Bees: Biology of a Superorganism*. Berlin: Springer, 2008.

Tavasli, Jusuf. *Dove i njihovi fadileti*. Sarajevo: Libris, 2013.

Tlili, Sarra. *Animals in the Qur'an*. New York: Cambridge University Press, 2012.

Trnka, Susanna. *One Blue Child: Asthma, Responsibility, and the Politics of Global Health*. Stanford, CA: Stanford University Press, 2017.

Tsing, Anna. "Blasted Landscapes (and the Gentle Arts of Mushroom Picking)." In *The Multispecies Salon*, edited by Eben Kirksey, 87–110. Durham, NC: Duke University Press, 2014.

———. "Unruly Edges: Mushrooms as Companion Species." *Environmental Humanities* 1, no. 1 (November 2012): 141–154. https://doi.org/10.1215/22011919-3610012.

Udruženje poljoprivrednika u Zeničko-dobojskom Kantonu. "Apel Vladi Federacije za hitno proglašavanje stanja elementarne nepogode." *Zenicablog*, August 5, 2017. https://www.zenicablog.com/apel-vladi-federacije-za-hitno-proglasavanje-stanja-elementarne-nepogode/.

Umeljić, Veroljub. *U svetu cveća i pčela: Atlas medonosnog bilja 1*. Kragujevac: Veroljub Umeljić, 2006.

UNDP Bosnia and Herzegovina. "Climate Change Adaptation and Low Emission Development Strategy for BiH, 2013." Research & Publications. Energy and Environment. January 9, 2014. https://www.ba.undp.org/content/bosnia_and_herzegovina/en/home/library/environment_energy/climate-change-adaptation-and-low-emission-development-strategy-.html.

———. "Third National Communication (TNC) and Second Biennial Update Report on Greenhouse Gas." Publications. Posted July 12, 2017. https://www.ba.undp.org/content/bosnia_and_herzegovina/en/home/library/environment_energy/tre_i-nacionalni-izvjetaj-bih.html.

UNFAO (Food and Agricultural Organization of the United Nations). "Imagine a World without Bees." UNFAO YouTube Channel. Video, 01:26. Posted May 18, 2018. https://www.youtube.com/watch?v=el-Z5tgyQXY.

UN Food and Agriculture Organization. "Chapter 6. Royal Jelly." Accessed January 2, 2018. https://www.fao.org/3/w0076e/w0076e16.htm.

van Dooren, Thom. *Flight Ways: Life and Loss at the Edge of Extinction*. New York: Columbia University Press, 2014.

Van Gelder, Linda. "It Is Time for Action on Climate Risk in the Balkans." The World Bank. Opinion. Posted September 17, 2018. https://www.worldbank.org/en/news/opinion/2018/09/17/it-is-time-for-action-on-climate-risk-in-the-balkans.

Velji, Jamel. *An Apocalyptic History of the Early Fatimid Empire (Edinburgh Studies in Islamic Apocalypticism and Eschatology)*. Edinburgh: Edinburgh University Press, 2016.

Wan, Derrick C., Stefanie L. Morgan, Andrew L. Spencley, Natasha Mariano, Erin Y. Chang, Gautam Shankar, Yunhai Luo, Ted H. Li, Dana Huh, Star K. Huynh, et al. "Honey Bee Royalactin Unlocks Conserved Pluripotency Pathway in Mammals." *Nature Communications* 9, no.1 (December 2018): 5078, https://doi.org/10.1038/s41467-018-06256-4.

Wang, Qing, Xinjian Xu, Xiangjie Zhu, Lin Chen, Shujing Zhou, Zachary Yong Huang, and Bingfeng Zhou. "Low-Temperature Stress during Capped Brood Stage Increases Pupa Mortality, Misorientation and Adult Mortality in Honey Bees." *PLoS One* 11, no. 5 (May 2015): e0154547. https://doi.org/10.1371/journal.pone.0154547.

Winston, Mark. *Bee Time: Lessons from the Hive*. Cambridge, MA: Harvard University Press, 2014.

Woyciechowski, Michal, and Dawid Moroń. "Life Expectancy and Onset of Foraging in the Honeybee (*Apis mellifera*)." *Insect Sociaux* 56, no. 2 (July 2009): 193–201. 10.1007/s00040-009-0012-6.

Yusuf, Hamza. "Death, Dying, and the Afterlife in the Quran." In *The Study Quran: A New Translation and Commentary*, edited by Seyyed Hossein Nasr, Caner K. Dagli, Maria Massi Dakake, Joseph E. B. Lumbard, and Mohammed Rustom, 1819–1865. New York: HarperCollins, 2015.

Zournazi, Mary, and Isabelle Stengers. "A Cosmopolitics—Risk, Hope, Change." In *Hope: New Philosophies for Change*, 244–272. New York: Routledge, 2003.

# INDEX

ākhir al-zaman, 20, 32, 43, 123
Al-Ghazālī, 70–72, 81, 161–162, 161, 164–165, 167, 172
angels, vii, ix, xi, 1–2, 5–8, 21, 39, 130, 132–135, 137, 161, 163, 167; 'Azrā'īl, 164; Isrāfīl, 158, 172
apiculture, x, 6, 16, 64, 76, 90, 115n2; in crisis, 103, 109
apitherapy, 45n12
apocalypse or apocalyptic, 3, 5–6, 21, 25, 34–35, 39, 44n2, 85, 87, 100, 109, 123, 143, 159, 160, 174–175, 179, 182; apocalypticism 3
artificial or supplemental feeding of bees, xi, 52, 53, 64, 81, 89, 102–103, 147

black locust, 30, 50, 51–53, 55, 57, 63, 66, 67, 68–69, 75, 76, 82, 89, 94, 104, 112, 125, 126, 127, 168

climate change, 44, 60, 63–64, 77, 96, 141, 146, 156n39, 159, 160; crisis, catastrophe, or disaster, 24, 59, 143, 148, 153n26; effects of, ix, x, 26; effects on honey ecologies, viii, 29, 52, 53, 64; effects on plants and pollinators, 28, 69, 81–82, 84n34, 87, 89, 91, 116n3, 145–146, 147, 156n40, 157n42, 179; and end times, 32, 43, 55, 100, 123–124; futures under, 2, 37, 122, 142, 143, 154n28, 175; policy, 144, 155n35; threats or risks, 3, 8, 122, 142; and UNDP, 155n36; and World Bank, 143
Čolić, Mustafa, xi, 21, 46n20, 46n23, 48n35, 72, 99, 100, 139, 188n14; on Imam Birgivi 21, 72, 139

disaster, 7–8, 23, 43, 57, 62, 97, 99, 122–123, 124–126, 141–143, 145, 154nn28–29, 155n35; quiet of, 34, 36
divine revelation, 20, 25, 26, 27, 48n35, 65, 97, 99, 100, 113, 121, 123, 128, 129–130, 136, 145, 149–150, 151, 161, 164, 166, 170, 173, 175, 176, 180, 181; to bees, 17–19, 33, 129; God's self-revelation or self-disclosure, 6, 23, 27, 99, 166, 181

ecology or ecologies, 2, 38, 47n29, 87, 97, 149; of honey, viii, 3, 28–32, 53; near-end ecologies, 144–145; of pollination, 2

end times or final times, 3, 19–24, 26, 32, 65, 66, 123

eschatology, 4, 21, 38, 39, 44, 159–160, 174, 182n4; eco-eschatology, 6, 23–24, 27–28

false springs, 29, 127

fasād, 6, 123–124, 128–29, 133, 137, 140, 151–152

foraging, 15, 17–18, 23, 37, 51, 52, 55, 61, 88, 95, 99, 104, 105, 116n3, 146; extreme weather effects on, 30–31, 89, 126; forage fronts, 37, 51–53, *54*, 69, 125; foragers, 27, 51, 56, 68, 75, 80, 90, 95, 107, 112, 116n3, 116n5, 126, 147; itineraries, 53, 68, 76

ghayb, 121–122, 130, 132, 134

hadith, vii, ix, x, 5, 17, 19, 20, 21, 26–27, 38, 44, 46n25, 66, 70, 71, 113, 166, 172, 173, 180; on planting, 85–86, 101, 109, 114

honey, 55, 57, 63, 67, 79, 90, 101, 102, 105, 107, 110, 112, 128, 137, 146, 149, 159, 168–169, 170; chances of, 61, 64, 69, 75, 148; flavors of, 52, 57, 67, 75, 78, 82, 127, 146, 169; flow, 2, 10, 51, 80; forecasts, 29, 63; as fruit of inspiration or revelation, 18, 26–27, 99, 121, 128, 136, 161, 181; as human nutrition and medicine, 6, 18, 86, 106, 115n2; importance for the hive, 14, 86, 95, 128, 149; strategies for catching or promoting the flow of, 3, 15, 28–29, 37, 53, 56, 64–65, 87, 94, 125; substitutes for, 87–90, 127, 128, 146; waning of, 2, 25, 52, 54, 72, 82, 87, 91, 103–104, 146

honeybees, x, xi, 23, 25, 53, 55, 67, 80, 94–95, 105–106, 109, 112, 123, 124, 126, 150, 159, 176, 182; climate futures of, 2, 91; community, society, or ummah, 16; decline or endangerment of, 26, 87–88, 116n3, 126, 161, 176–180, 183n40; and hive affairs, 31, 53, 57, 88, 95, 126, 127, 128, 152n7; and inspiration or revelation, 6, 16–18, 33, 99, 129–130, 181; invocation of, 9, 170–171; resilience of, 147–148, 157n42; swarms or swarming, 10–15, 181; world without, 177–180

honeydew, 55, 57, 75, 76, 82, 90

Ibn al-ʿArabī, xi, 74, 145, 181

indigo bush or false indigo, 55, 57, 62, 76, 111, 135, 168

insān or insan, 23, 99, 109, 132, 137, 139, 151, 177

invasive species, 28, 51, 110

Jerusalem thorn, 66, *73*, 74–75

jinn, 39, 72, 109, 130–131, 132, 133, 134–137, 139, 176

kijamet. See qiyāmah or kijamet

linden: 55, 57, 63, 76, 82, 87, 89, 94, 104, 109, 112, 126, 146, 147

meadows, 69, 76, 77, 78, 82, 93, 101, 102, 104, 105, 126, 146, 147, 169

metaphysics or the metaphysical, 5, 22, 23, 25, 39, 48n29, 54, 60, 61, 65, 72, 80–81, 117n6, 160, 175

nafs, 46n24, 108–109, 121, 130, 135, 136, 138, 151

Nasr, Seyyed Hossein, 46n25

plants, planting, or gardening: 3, 29, 46n25, 86–87, 91, 93–96, 100–101, 106, 109–115, 132, 145

pollen, 2–3, 28, 30, 32, 36, 37, 51, 53, 55, 62, 81, 82, 87, 91, 103, 148; in bee's diet, 57, 89, 90, 94, 95, 102, 112, 116n5, 126–127, 128, 149; in human medicine and diet, 106; substitutes for, 67, 88, 102, 157n44

pollination, 2, 28, 81, 126, 159, 180, 183n38; crisis of, 178, 184n40

prayers or supplications, 10, 17, 54, 55, 61, 70, 71, 80, 130, 136, 145, 150, 167; the devil's, 71, 131, 133

Prophet, the, vii, x, 4, 18, 21, 46n24, 71, 83n14, 85, 99, 145, 150, 153n18, 161, 180, 181, 183n26, 183n28

propolis, 95, 106, 118n16

qiyāmah or kijamet, 1, 8, 34, 35, 99, 109, 123, 158–159, 161–164, 168, 174, 176, 177

royal jelly or bee mother's milk, 106, 116n4, 118n16

savory, 77, 80, 84n29

Shaitan, the devil, or Iblīs, 71, 124, 130, 133, 134, 136–140, 145, 151, 152

tawbah, 70–71, 83n13

venom, 106, 118n17, 149, 157n43

weather, 5, 36, 52, 55, 64, 70, 78, 80; catastrophic, extreme, or strange, 2, 7–8, 30, 35, 37, 121, 122, 123–124, 126, 143, 144, 146, 154n26, 156n39; effects of climate change on bees and plants, 6, 9, 28, 53, 63–65, 68, 79, 81, 84n34, 89, 91, 96, 126–27, 146–48, 149; Lucifer heatwave, 40–41, 122; May 2014 storm, 7–9, 124–126, 127, 142, 155n29; and nectar flow, 29, 37, 44, 53, 63; and seasons, 30

**Larisa Jašarević** is an anthropologist. A fellow with the Independent Social Research Foundation and a visiting scholar at the Max Planck Institute for the History of Science, she lives and works by her apiary in northeastern Bosnia. Larisa is an independent scholar and has previously taught at the University of Chicago. She is author of *Health and Wealth on the Bosnian Market: Intimate Debt*.

## For Indiana University Press

Tony Brewer, *Artist and Book Designer*
Brian Carroll, *Rights Manager*
Allison Chaplin, *Acquisitions Editor*
Sophia Hebert, *Assistant Acquisitions Editor*
Brenna Hosman, *Production Coordinator*
Katie Huggins, *Production Manager*
Nancy Lightfoot, *Project Editor and Manager*
Dan Pyle, *Online Publishing Manager*
Samantha Heffner, *Marketing and Publicity Manager*
Leyla Salamova, *Senior Artist and Book Designer*

Printed and bound by CPI Group (UK) Ltd, Croydon, CR0 4YY
04/03/2024

14461303-0001